Science Policy: New Mechanisms for Scientific Collaboration between East and West

NATO ASI Series

Advanced Science Institutes Series

A Series presenting the results of activities sponsored by the NATO Science Committee, which aims at the dissemination of advanced scientific and technological knowledge, with a view to strengthening links between scientific communities.

The Series is published by an international board of publishers in conjunction with the NATO Scientific Affairs Division

A Life Sciences **B Physics**	Plenum Publishing Corporation London and New York
C Mathematical and Physical Sciences **D Behavioural and Social Sciences** **E Applied Sciences**	Kluwer Academic Publishers Dordrecht, Boston and London
F Computer and Systems Sciences **G Ecological Sciences** **H Cell Biology** **I Global Environmental Change**	Springer-Verlag Berlin, Heidelberg, New York, London, Paris and Tokyo

PARTNERSHIP SUB-SERIES

1. Disarmament Technologies	Kluwer Academic Publishers
2. Environment	Springer-Verlag
3. High Technology	Kluwer Academic Publishers
4. Science and Technology Policy	Kluwer Academic Publishers
5. Computer Networking	Kluwer Academic Publishers

The Partnership Sub-Series incorporates activities undertaken in collaboration with NATO's Cooperation Partners, the countries of the CIS and Central and Eastern Europe, in Priority Areas of concern to those countries.

NATO-PCO-DATA BASE

The electronic index to the NATO ASI Series provides full bibliographical references (with keywords and/or abstracts) to more than 30000 contributions from international scientists published in all sections of the NATO ASI Series.
Access to the NATO-PCO-DATA BASE is possible in two ways:

– via online FILE 128 (NATO-PCO-DATA BASE) hosted by ESRIN, Via Galileo Galilei, I-00044 Frascati, Italy.

– via CD-ROM "NATO-PCO-DATA BASE" with user-friendly retrieval software in English, French and German (© WTV GmbH and DATAWARE Technologies Inc. 1989).

The CD-ROM can be ordered through any member of the Board of Publishers or through NATO-PCO, Overijse, Belgium.

4. Science and Technology Policy – Vol. 1

Science Policy: New Mechanisms for Scientific Collaboration between East and West

edited by

Valentin A. Koptyug

Siberian Branch of the Russian Academy of Sciences,
Novosibirsk, Russia

and

Jean Klerkx

Department of Geology and Mineralogy,
Royal Museum of Central Africa,
Tervuren, Belgium

Springer Science+Business Media, B.V.

Proceedings of the NATO Advanced Research Workshop on
Science Policy: New Mechanisms for Scientific Collaboration between East and West
Novosibirsk, Russia
November 22–25, 1993

A C.I.P. Catalogue record for this book is available from the Library of Congress.

ISBN 978-94-010-4074-7 ISBN 978-94-011-0165-3 (eBook)
DOI 10.1007/978-94-011-0165-3

Printed on acid-free paper

CONTENTS

Preface ix

List of Participants xi

List of Observers xxiii

List of Contributors xxvii

I. Current Problems and Prospects of International Collaboration

New Russian Problems of International Scientific Collaboration 3
B.G. Saltykov

NATO International Scientific Exchange Programs 9
A.H. Jubier

A Network of Research Centres in Siberia, as a Basis for Multi-Disciplinary International Scientific Collaboration 13
V.A. Koptyug

Science Policy: New Mechanisms for Scientific Collaboration Between East and West - UNESCO's Role and Possible Contributions 19
F.W. Eder

Research Programs of the Council of Europe and Possible Ways of Collaboration with Scientists from Eastern Countries 27
J.P. Massue

East-West Scientific Collaboration - Some Ideas from the Laboratory
M. André 31

II. Experience of the International Research Centres' Activities and Some Results of Interaction between Scientists in Priority and Multidisciplinary Problems

II.1 Environmental Problems

International Cooperation in Boreal Forests Study of Siberia
E.A. Vaganov, A.S. Isaev and S. Nilsson 37

Global Change, the Global Boreal Forest, and Fire: Search for New Strategies in Science Policies and Research Mechanisms
J.G. Goldammer and V.V. Furiaev 45

Man-made Closed Ecological Systems: Scientific and Applied Significance, and Prospects for International Collaboration
J.I. Gitelson, I.I. Lisovsky and V.E. Rygalov 63

Genetic Effects of Radiation and other Man-made Pollution in Siberia
V.K. Shumny, N.A. Kolchanov, N.N. Dygalo, V.G. Matveeva, L.N. Ivanova, M.I. Voevoda,A.G. Romashchenko, E.I. Karakin, I.I. Kiknadze, E.B. Budashkina, S.E. Peltek 69

Integrated Sustainable Development Methodology
St. Bonfa 81

II.2. Interdisciplinary Studies of Lake Baikal

Baikal International Centre for Ecological Research: Principles of Organization and First Scientific Results
M.A. Grachev 89

A Brief Report of Japanese Activity for BICER Projects in 1993
T. Kawai 101

Baikal Drilling Project: First Results and Prospects for Future Studies
M.I. Kuzmin, N.A. Logachev, M.A. Grachev, P.P. Hearn, D.F. Williams, Sh. Horie and T. Kawai 107

The CASIMIR Project: a Joint Research Project on Rift Basin Geology between Siberia and Belgium
J. Klerkx, N.A. Logachev V.D. Ermikov and K.G. Levi 117

II.3. Natural Sciences and Technological Applications

International Cooperation in the Framework of the Siberian Synchrotron Radiation Centre
G.N. Kulipanov 129

International Centre for Catalyst Characterization and Testing: Two Years of Activity
K.I. Zamaraev, J.M. Thomas and J.H. Block 141

Towards Sustainable Development: Contributions of the Boreskov Institute of Catalysis
H.J. Lovink 153

Material Science Program of the Siberian Branch of the Russian Academy of Sciences
F.A. Kuznetsov 159

The Conversion of Basic Chemical Technologies into the Solid State Processes
N.Z. Lyakhov and V.V. Boldyrev 163

Space and Environmental Research at the SB RAS Facilities
G.Ya. Smolkov and G.A. Zherebtsov 167

Space Contamination and Ecological Problems
V.M. Fomin, A.M. Kharitonov, G.A. Sapozhnikov and E.I. Grinberg 177

Cooperation in Environmental Flow Problems of Aerospace Science
G.E.A. Meier 185

Longterm Cooperation of SB RAS and Bruker-Spectrospin Group of Companies: Results and Prospects
R.Z. Sagdeev, A.V. Podoplelov and G. Laukien 193

II.4. Socio-Economic and Humanitarian Studies

International Cooperation in Studies of Economic Development of Siberia
V.E. Seliverstov and S.A. Savey 199

The Baikal Region Social Ecology Issues
A.K. Tulokhonov, B.O. Gomboev and T.B. Bardakhanova 209

Perspectives and Preliminary Results of Archaeological Investigations of the South-Western Altai (the Ukok Plateau)
V.I. Molodin 215

Some Aspects of Collaboration between Belgium Royal Museums of Art and History and Russian Colleagues in Archaeological and Art Historian Matters
F. Van Noten 223

III. Information Support and Education

Information Service for Scientific Programs in the Siberian Branch of the Russian Academy of Sciences
B.S. Elepov, L.K. Bobrov, O.L. Lavrik, S.V. Morosov and E.B. Soboleva 227

State and Perspectives of Cooperation of the Institutes of Siberian Branch of Russian Academy of Sciences with Western Specialists in the Field of Geophysical Informatics
A.S. Alekseev 235

Interaction of Education and Science in SB RAS and International Collaboration in Education and Training of Junior Scientists
Yu.L. Ershov 247

Statement 253

PREFACE

On the eve of the XXI century our civilization is colliding with the next global problem, demanding a global solution. It is becoming more and more clear that the future of the world will be impossible without scientific support, analysis, prediction, and the use of scientific achievements. The immensity of these problems, and the limited financial resources which society may spend on their research and solution, urgently demand the selection of priority areas of research and scientists' combined efforts on a global scale.

The new geopolitical situation has uncovered Russia's and the former Soviet Union's scientific potential to the rest of the world community. This situation has given rise to wider opportunities for involvement in work on international projects and programs. The above-mentioned considerations impelled the Siberian Branch of the Russian Academy of Sciences to organize a workshop, under NATO (Scientific and Environmental Affairs Division) sponsorship, "Science Policy: New Mechanisms for Scientific Collaboration Between East and West". This Workshop was held in Novosibirsk in November 22-25, 1993.

The Workshop participants, from 17 countries, represented research institutes, universities, as well as NATO's scientific bodies, UNESCO, the Council of Europe, and other international organizations. The range of discussed problems included current problems, as well as prospects of international collaboration of scientists from both the East and the West. The scope of discussion also included interdisciplinary priority programs demanding the combined efforts of scientists from many countries, the experience in organizing the work of international research centres in Siberia as a new and promising form of international collaboration, some of their scientific results, and a suggestion of the direction in which collaboration should move. By the way, this explains the diverse character of the papers presented in this volume.

The Workshop's joint Statement is given at the end of the volume. It specifically addresses priority fields of science in which it was found expedient to develop joint multidisciplinary research. This research could encompass rational use of natural resources and regional sustainable development, as well as the monitoring of the current biosphere's ecosystem state and the risks of natural and anthropogenic hazards. The research might also broach the creation of advanced technologies and materials. The list of priorities included Siberia's economic, social, and humanitarian problems, as well as development of information systems for quick exchange of scientific data. There was a recognized need for wider access for young scientists' practice in partnership laboratories while on probation. This would include Siberian research centers.

The Workshop participants addressed the following organizations in their bid for attention for the Statement and actions to support East-West scientific collaboration: UNESCO, the International Council of Scientific Unions, NATO Science Committee, the Council of Europe, the Commission of the European Community, and the governments of the Commonwealth of Independent States (CIS).

The organisers gratefully acknowlegde the financial support of the Scientific Affairs Division of NATO for the organisation of the Workshop.

LIST OF PARTICIPANTS

Prof. Anatoly S. ALEKSEEV
Siberian International Centre of
New Information Technologies
Computing Centre of SB RAS
Director
6 Lavrentiev Ave.
630090 Novosibirsk
Tel: 7 (383-2) 355-650
Fax: 7 (383-2) 354-269
E-mail: aleks@comcen.nsk.su
RUSSIA

Dr. Marcel ANDRE
Centre d'Etudes Nucléaires de Cadarache
Conseiller Scientifique
13108 Saint-Paul-Lez-Durance Cedex
Tel: 33 (42) 257-578
Fax: 33 (42) 254-656
E-mail: thiery@orstom.orstom.fr
FRANCE

Dr. Michael ANDRISKE
Ruhr-University of Bochum
Comparative Endocrinology
Research Section, Faculty of Biology
Project Scientist/CEBAS Aquatade
D-44780 Bochum
Tel: 49 (234) 700-4332 or 700-2846
Fax: 49 (234) 709-4189
GERMANY

Dr. Jean-Claude BIDET
CIFEG
Head of Documentation PANGIS
coordinator
Avenue de Concyr B.P. 6517
F-45065 Orl ans Cedex 2
Tel: 33 (38) 643-367
Fax: 33 (38) 643-472
Telex: 780258 ATTN. CIFEG
E-mail: D.NGVYEN@orleans.brgm.fr
FRANCE

Prof. Vladimir V. BOLDYREV
Institute of Solid State Chemistry of SB
RAS
Director
18 Derzhavin Str.
630091 Novosibirsk
Tel: 7 (383-2) 200-423
Fax: 7 (383-2) 200-969
E-mail: boldyrev@solid.nsk.su
RUSSIA

Dr. Stefano BONFA
Environmental Research
Association Mediterranean
Programme Coordinator
15 Via Tiberio Imperatore
00145 Roma
Tel: 39 (6) 512-6690
Fax: 39 (6) 512-8901
E-mail: maurizi@irmias.ias.fra.cnr.it
ITALY

Mrs. Claire BRONE
Royal Museum of Central Africa
Department of Geology
Administrative Secretary
B-3080 Tervuren
Tel: 32 (2) 769-5430
Fax: 32 (2) 769-5432
Telex: 27048 BUGEO B
E-mail: geomrac@vub.ac.be
BELGIUM

Dr. Victor V. BUGROVSKY
Uvs-Nuur International Centre for
Biospheric Research
Project Leader
667000 Kyzyl, Box 107
Tel: 7 (394-22) 37-160
RUSSIA

Prof. CHOI Mong-Lyong
Seoul National University
Dept. of Archaeology and Art History
College of Humanities
Seoul 151-542
Tel: 82 (2) 880-6212
Fax: 82 (2) 851-5244
KOREA

Prof. Anatoly P. DEREVYANKO
Altai International Centre for
Humanitarian and Biospheric Research
Scientific Leader of the Humanitarian Part
Institute of Archaeology and
Ethnography of SB RAS
Director
17 Lavrentiev Ave.
630090 Novosibirsk
Tel: 7 (383-2) 350-537
Fax: 7 (383-2) 357-791
RUSSIA

Prof. Nikolai L. DOBRETSOV
Siberian Branch of RAS
Vice-Chairman
17 Lavrentiev Ave.
630090 Novosibirsk
Tel: 7 (383-2) 350-566
Fax: 7 (383-2) 354-846
Telex: 133128 POISK SU
E-mail: dobr@hq.soan.nsk.su
RUSSIA

Prof. Vyacheslav P. DOBRITCA
State University of Kazakhstan
13 Lenin Str.
480100 Alma-Ata
Tel: 7 (327-2) 613-769
Fax: 7 (327-2) 613-050
KAZAKHSTAN

Dr. Nikolay N. DYGALO
Institute of Cytology and Genetics of SB RAS
Chief Researcher
10 Lavrentiev Ave.
630090 Novosibirsk
Tel: 7 (383-2) 351-250
Fax: 7 (383-2) 356-558
RUSSIA

Dr. Franz W. EDER
UNESCO, Division of Earth Sciences
Director
1 rue Miollis
F-75732 Paris Cedex 15
Tel: 33 (1) 4568-4115
Fax: 33 (1) 4306-122
FRANCE

Dr. Boris S. ELEPOV
State Public Scientific Library of SB RAS
Director
15 Voshod Str.
630200 Novosibirsk
Tel: 7 (383-2) 661-860
Fax: 7 (383-2) 660-308
Telex: 133220 LIBRO SU
E-mail: elepov@libr.nsk.su
RUSSIA

Prof. Jaeque (Jody) EMEL
Clark University
Graduate School of Geography
950 Main Street Worcester,
Massachusetts 01610-1477
Tel: 1 (508) 793-7317
Fax: 1 (508) 793-8881
E-mail: jemel@vax.clarku.edu
USA

Dr. Valery D. ERMIKOV
Presidium of SB RAS
Chief of Department
17 Lavrentiev Ave.
630090 Novosibirsk
Tel: 7 (383-2) 356-619
Fax: 7 (383-2) 354-846
E-mail: vde@hq.soan.nsk.su
RUSSIA

Prof. Yuri L. ERSHOV
Novosibirsk State University
Past Rector
2 Pirogov Str.
630090 Novosibirsk
Tel: 7 (383-2) 357-808
Fax: 7 (383-2) 356-237
E-mail: erchov@mioo.cnit.nsk.su
RUSSIA

Prof. Vasily M. FOMIN
Siberian Centre of Aerophysical Studies
Institute of Theoretical and
Applied Mechanics of SB RAS
Director
4/1 Institutskaya Str.
630090 Novosibirsk
Tel: 7 (383-2) 353-533
Fax: 7 (383-2) 352-268
Telex: 133128 POISK SU
E-mail: fomin@icas.nsk.su
RUSSIA

Prof. Josef I. GITELSON
International Centre for
Closed Ecological Systems
Institute of Biophysics of SB RAS
Director
660036 Krasnoyarsk, Academgorodok
Tel: 7 (391-2) 434-623
Fax: 7 (391-2) 222-556
Telex: 288139 MOST SU
E-mail: jgit@ibp.krasnoyarsk.su
RUSSIA

Dr. Johann G. GOLDAMMER
Max-Planck Institut for Chemistry
Leader of the Fire Ecology Research Group
D-79085 Freiburg
Tel: 49 (761) 808-011
Fax: 49 (761) 808-012
Telex: 4187674 MPCH D
E-mail: jggold@sun1.ruf.uni-freiburg.de
GERMANY

Dr. Sergey S. GONCHAROV
Siberian International Centre of
New Information Technologies
Executive Director
6 Lavrentiev Ave.
630090 Novosibirsk
Tel: 7 (383-2) 354-462 or 356-237
RUSSIA

Prof. Gennady I. GRITSKO
Kemerovo Scientific Centre of SB RAS
Chairman of Presidium
21 Rukavishnikov Str.
650610 Kemerovo
Tel: 7 (384-2) 281-433
Telex: 215113 NAUKA SU
RUSSIA

Prof. Michael A. GRACHEV
Baikal International Centre for
Ecological Research
Limnological Institute of SB RAS
Managing Director
3 Ulan-Batorskaya Str.
664033 Irkutsk
Tel: 7 (395-2) 460-504
Fax: 7 (095) 420-2106 (Moscow)
E-mail: grachev@lin.irkutsk.su
RUSSIA

Prof. HWANG Yong-Hoon
Kyung Hee University
Institute of Archaeology and Art History
Seoul 130-701
Tel: 82 (2) 961-0070 or 965-9483
Fax: 82 (2) 965-8004
KOREA

Dr. Victor V. INOZEMTSEV
Coordinating Unit for
EC Technical Assistance
Expert in Nuclear Safety &
Military Conversion
Slavianskaja Hotel, Office 5032
2 Berezhkovskaja Nab.
121059 Moscow
Tel: 7 (095) 941-8653
Fax: 7 (095) 941-8643
RUSSIA

Prof. Aleksandr S. ISAEV
International Forest Institute
Director
69 Novotcheremushkinskaja Str.
117418 Moscow
Tel: 7 (095) 332-8652
Fax: 7 (095) 332-6020
RUSSIA

Dr. Aleksandr S. IVANOV
Tomsk Scientific Centre of SB RAS
Chairman Assistant
2 Academitchesky Ave.
634055 Tomsk
Tel: 7 (382-2) 259-276
E-mail: tnc@elecard.tomsk.su
RUSSIA

Dr. Alain H. JUBIER
NATO, Scientific Affairs Division
Program Director
B-1110 Brussels
Tel: 32 (2) 728-4228
Fax: 32 (2) 728-4232
BELGIUM

Dr. Marina KAZAKEVITCH
ERAM
Moshav Gealia 78
Rehovot 76885
Tel: 972 (8) 470-374
Fax: 972 (8) 468-620
E-mail: diskin@wisdom.weizmann.ac.il
ISRAEL

Prof. Anatoly M. KHARITONOV
International Center of Aerophysical Studies
Executive Director
4/1 Institutskaya Str.
630090 Novosibirsk
Tel: 7 (383-2) 350-778
Fax: 7 (383-2) 352-268
E-mail: khar@icas.nsk.su
RUSSIA

Dr. Vladimir A. KHRIPACH
Academy of Sciences of Belarus
Deputy Secretary General
66 Francisk Ckorina Ave.
220072 Minsk
Tel: 7 (017-2) 394-805
Fax: 7 (017-2) 393-163
BELARUS

Prof. J. KLERKX
Royal Museum of Central Africa
Head of Department of Geology
13 Leuvensesteenweg
B-3080 Tervuren
Tel: 32 (2) 769-5426
Fax: 32 (2) 769-5432
Telex: 27048 BUGEO B
E-mail: geomrac@vub.ac.be
BELGIUM

Dr. Helmut KLUTER
Geographisches Institut der
Justus-Liebig-Universität Giessen
Privatdocent
Senckenbergstrasse 1
D-35390 Giessen
Tel: 49 (641) 702-8265
Fax: 49 (641) 702-8211
GERMANY

Dr. Nikolay A. KOLCHANOV
Institute of Cytology and Genetics of SB RAS
Deputy director
10 Lavrentiev Ave.
630090 Novosibirsk
Tel: 7 (383-2) 353-335
Fax: 7 (383-2) 356-558
E-mail: root@cgi.nsk.su
RUSSIA

Dr. Vladimir N. KOLYAKIN
Ministry of Science and Technical Policy of the Russian Federation
Chief Specialist
11 Tverskaya Str.
103905 Moscow
Tel: 7 (095) 229-5911
Fax: 7 (095) 230-2660
RUSSIA

Dr. Edward G. KONNIKOV
Geology Institute of Buryatia of SB RAS
Director
6a Sahjanov Str.
670042 Ulan-Ude
Tel: 7 (301-22) 30-955
E-mail: root@bien.buriatia.su
RUSSIA

Prof. Valentin A. KOPTYUG
Siberian Branch of RAS
Chairman
17 Lavrentiev Ave.
630090 Novosibirsk
Tel: 7 (383-2) 350-567
Fax: 7 (383-2) 354-846
7 (095) 938-2097
Telex: 133128 POISK SU
E-mail: vak@hg.soan.nsk.su
RUSSIA

Prof. Valery V. KULESHOV
Institute of Economics and Industrial Engineering of SB RAS
Director
17 Lavrentiev Ave.
630090 Novosibirsk
Tel: 7 (383-2) 350-536
E-mail: uno-ie@sinor.nsk.su
RUSSIA

Prof. Igor Yu. KOROPATCHINSKY
Uvs-Nuur International Centre for Biospheric Research
Central Siberian Botanical Gardens of SB RAS
Director
101 Zolotodolinskaya Str.
630090 Novosibirsk
Tel: 7 (383-2) 354-101
Fax: 7 (383-2) 354-986
RUSSIA

Dr. Vladimir A. KRUTIKOV
Tomsk Scientific Centre of SB RAS
Chairman of Presidium
2 Academitchesky Ave.
634055 Tomsk
Tel: 7 (382-2) 259-309
Fax: 7 (382-2) 259-713
Telex: 612510 LIDAR SU
E-mail: tnc@elecard.tomsk.su
RUSSIA

Dr. Gennady N. KULIPANOV
Siberian Synchrotron Radiation Centre
Budker Institute of
Nuclear Physics of SB RAS
Deputy Director
11 Lavrentiev Ave.
630090 Novosibirsk
Tel: 7 (383-2) 359-498
Fax: 7 (383-2) 352-163
Telex: 133819 KORUND SU
E-mail: kulipanov@inp.nsk.su
RUSSIA

Dr. Svetlana S. KURBATSKAYA
Uvs-Nuur International Centre for
Biospheric Research
Executive Director
667000 Kyzyl, Box 107
Tel: 7 (394-22) 37-160
RUSSIA

Prof. Michael I. KUZMIN
Vinogradov Institute
of Geochemistry of SB RAS
Director
1a Favorsky Str.
664033 Irkutsk
Tel: 7 (395-2) 460-500
Fax: 7 (395-2) 462-952
E-mail: mikuzmin@igc.irkutsk.su
RUSSIA

Prof. Fedor A. KUZNETSOV
Institute of Inorganic Chemistry of SB RAS
Director
3 Lavrentiev Ave.
630090 Novosibirsk
Tel: 7 (383-2) 355-950
Fax: 7 (383-2) 355-960
Telex: 133226 VODA SU
E-mail: fk@che.nsk.su
RUSSIA

Dr. Vladimir I. LEBEDEV
Tuva Interdisciplinary
Department of SB RAS
Director
117a International Str.
667007 Kyzyl
Tel: 7 (394-22) 25-006
Fax: 7 (394-22) 32-578
RUSSIA

Dr. Kirill G. LEVI
International Centre for Active Tectonics
and Natural Disasters
Institute of the Earth's Crust of SB RAS
Deputy Director
128 Lermontov Str.
664033 Irkutsk
Tel: 7 (395-2) 461-562
E-mail: levi@cora.irkutsk.su
RUSSIA

Dr. Françoise LEVIE
SOFIDOC S.A.
Director
Rue Rasson, 19-21
B-1040 Brussels
Tel: 32 (2) 736-5437
Fax: 32 (2) 736-5239
BELGIUM

Prof. Nikolay A. LOGACHEV
International Centre for Active Tectonics
and Natural Disasters
Institute of the Earth's Crust of SB RAS
Director
128 Lermontov Str.
664033 Irkutsk
Tel: 7 (395-2) 464-000
E-mail: log@cora.irkutsk.su
RUSSIA

Prof. Hermanus J. LOVINK
Akzo Catalysts
Senior Technical Associate
Stationsplein 4, P.O. Box 247
3800 AE Amersfoort
Tel: 31 (033) 676-317
Fax: 31 (033) 676-154
Telex: 79276 ACAF NL
The NETHERLANDS

Prof. Nikolay Z. LYAKHOV
Institute of Solid State Chemistry of SB RAS
Deputy Director
18 Derzhavin Str.
630091 Novosibirsk
Tel: 7 (383-2) 200-949
Fax: 7 (383-2) 200-969
E-mail: boldyrev@solid.nsk.su
RUSSIA

Dr. Arkady A. MALTSEV
IIASA
Deputy Director
Schlossplatz 1, Laxenburg
Fax: 43 (223) 671-313
Telex: 079137 IIASA A
E-mail: ramirez@iiasa.ac.at
AUSTRIA

Dr. Jean-Pierre MASSUE
Council of Europe
European Federation of Scientific Networks
Secretary General
B.P. 431/R6, F-67006 Strasbourg Cedex 2
Tel: 33 (88) 412-614, 412-615
Fax: 33 (88) 412-787
FRANCE

Prof. Dr. Gerd E.A. MEIER
DLR-Institute for Fluid Mechanics
Director
Bunsenstrasse 10, 37073 Göttingen
Tel: 49 (551) 709-2177
Fax: 49 (551) 709-2889
Telex: 96839 DFVGOE D
E-mail: sigfried@spock.es.go.dkr.de
GERMANY

Prof. Valery L. MIRONOV
Altai State University
Rector
66 Dimitrov Str.
656099 Barnaul
Tel: 7 (385-2) 222-881
Fax: 7 (385-2) 222-875
E-mail: vup@mezon.altai.su
RUSSIA

Prof. Vyacheslav I. MOLODIN
Altai International Centre for Humanitarian and Biospheric Research
Institute of Archaeology and Ethnography of SB RAS
Deputy Director
17 Lavrentiev Ave.
630090 Novosibirsk
Tel: 7 (383-2) 357-241
Fax: 7 (383-2) 357-791
RUSSIA

Prof. Vladimir E. NAKORYAKOV
Institute of Thermal Physics of SB RAS
Director
1 Lavrentiev Ave.
630090 Novosibirsk
Tel: 7 (383-2) 355-552
Fax: 7 (383-2) 357-880
E-mail: root@otani.thermo.nsk.su
RUSSIA

Prof. Sten NILSSON
IIASA
Project Leader
Schlossplatz 1, Laxenburg
Tel: 43 (223) 671-521 Ext. 229
Fax: 43 (223) 671-313
Telex: 079137 IIASA A
E-mail: nilsson@iiasa.ac.at
AUSTRIA

Dr. Leonid P. ORLOV
Government of Republic of Belarus
Science and New Technology Department
Chief Expert
Government House
220010 Minsk
Tel: 7 (017-2) 296-209
Fax: 7 (017-2) 296-665
BELARUS

Dr. Valentin N. PARMON
Boreskov Institute of Catalysis of SB RAS
Deputy Director
3 Lavrentiev Ave.
630090 Novosibirsk
Tel: 7 (383-2) 352-269
Fax: 7 (383-2) 355-766
E-mail: root@catalysis.nsk.su
RUSSIA

Dr. Aleksey V. PODOPLELOV
Siberian Tomography Centre
Deputy Director
3 Institutskaya Str.
630090 Novosibirsk
Tel: 7 (383-2) 353-550
Fax: 7 (383-2) 352-366
Telex: 133148 KING SU
E-mail: root@tomo.nsk.su
RUSSIA

Prof. Igor V. POTTOSIN
Institute of Information Systems of SB RAS
Director
6 Lavrentiev Ave.
630090 Novosibirsk
Tel: 7 (383-2) 355-652
Fax: 7 (383-2) 353-494
E-mail: viknst@isi.itfs.nsk.su
RUSSIA

Dr. Vadim E. RYGALOV
International Centre for
Closed Ecological Systems
Executive Director
660036 Krasnoyarsk, Academgorodok
Tel: 7 (391-2) 434-623
Fax: 7 (391-2) 222-556
Telex: 288144 KIF SU
E-mail: abrek@ibp.krasnoyarsk.su
RUSSIA

Prof. Renad Z. SAGDEEV
Siberian Tomography Centre
Director
3 Institutskaya Str.
630090 Novosibirsk
Tel: 7 (383-2) 352-366
Fax: 7 (383-2) 352-366
Telex: 133148 KING SU
E-mail: root@tomo.nsk.su
RUSSIA

Dr. Boris G. SALTYKOV
Minister of Science and Technical Policy of
the Russian Federation
11 Tverskaya Str.
103905 Moscow
Tel: 7 (095) 229-2501
Moskow
RUSSIA

Prof. Lev S. SANDAHCHIEV
Institute for Molecular Biology
General Director
Koltsovo
633159 Novosibirsk
Tel: 7 (383-2) 610-140
RUSSIA

Prof. Suzanne A. SAVEY
Université Paul Valéry
Géographie et Aménagement du Territoire
Route de Mende, B.P. 5043
34032 Montpellier Cedex 1
Tel: 33 (67) 142-120 or 272-980
Fax: 33 (67) 142-332
FRANCE

Dr. Vyacheslav E. SELIVERSTOV
Siberian International Centre for
Regional Studies
Executive Director
17 Lavrentiev Ave.
630090 Novosibirsk
Tel: 7 (383-2) 353-954
Fax: 7 (383-2) 355-580
E-mail: sel@sicrs.nsk.su
RUSSIA

Dr. Vasily F. SHABANOV
Krasnoyarsk Scientific Centre of SB RAS
Chairman of Presidium
660036 Krasnoyarsk, Academgorodok
Tel: 7 (391-2) 434-512
E-mail: prksc@prksc.krasnoyarsk.su
RUSSIA

Dr. Valery A. SHAPTSEV
Institute of Information Technologies and
Applied Mathematics of SB RAS
Director
28 Andrianov Str.
644077 Omsk
Tel: 7 (381-2) 653-727
E-mail: iitpm@intekh.omsk.su
RUSSIA

Prof. Yurii I. SHOKIN
Presidium of SB RAS
General Science Secretary
17 Lavrentiev Ave.
630090 Novosibirsk
Tel: 7 (383-2) 353-345
Fax: 7 (383-2) 353-647
Telex: 133128 POISK SU
E-mail: shokin@hq.soan.nsk.su
RUSSIA

Prof. Vladimir K. SHUMNY
Altai International Centre for Humanitarian
and Biospheric Research
Scientific Leader of Biospheric Research
Institute of Cytology and Genetics of SB
RAS
Director
10 Lavrentiev Ave.
630090 Novosibirsk
Tel: 7 (383-2) 354-350
Fax: 7 (383-2) 356-558
E-mail: root@cgi.nsk.su
RUSSIA

Dr. Gennady Ya. SMOLKOV
Institute of Solar-Terrestrial
Physics of SB RAS
Deputy Director
664033 Irkutsk, Box 4026
Tel: 7 (395-2) 463-791
Fax: 7 (395-2) 462-557
E-mail: root@icc.ccsoan.irkutsk.su
RUSSIA

Prof. Nikolai V. SOBOLEV
Institute of Mineralogy and
Petrography of SB RAS
Director
3 Universitetsky Ave.
630090 Novosibirsk
Tel: 7 (383-2) 354-655
Fax: 7 (383-2) 352-692
Telex: 133123 KORA SU
E-mail: eclogite@diamond.nsk.su
RUSSIA

Dr. Michael P. SVESHNIKOV
Presidium of Urals Branch of RAS
Head of Department
91 Pervomajskaja Str.
620219 Ekaterinburg
Tel: 7 (343-2) 493-095
Fax: 7 (343-2) 221-546
RUSSIA

Prof. Konstantin K. SVITASHEV
Institute of Semiconductor
Physics of SB RAS
Director
13 Lavrentiev Ave.
630090 Novosibirsk
Tel: 7 (383-2) 354-250
Fax: 7 (383-2) 354-265
E-mail: help-1@isph.nsk.su
RUSSIA

Dr. Alexandr V. TRAPEZNIKOV
Institute of Plant and Animal Ecology
Urals Branch of RAS
Head of Department
Ekaterinburg Region
624051 Zarechny
Tel: 7 (343-77) 32-070
RUSSIA

Dr. Arnold K. TULOKHONOV
International Scientific Centre of Socio-Ecological Problems of the Lake Baikal Basin
Baikal Institute of Nature Management of SB RAS
Director
8 Sahyanova Str.
670042 Ulan-Ude
Tel: 7 (301-22) 38-035
Fax: 7 (301-22) 63-444
Telex: 219117 LOTOS SU
E-mail: root@bien.buriatia.su
RUSSIA

Prof. Francis Van NOTEN
Royal Museums of Art and History
Director
10 Jubelpark
B-1040 Brussels
Tel: 32 (2) 741-7200
Fax: 32 (2) 733-7735
BELGIUM

Prof. Oleg F. VASILIEV
Institute for Water and
Environmental Problems of SB RAS
Director
105 Papanintsev Str.
656099 Barnaul
Tel: 7 (385-2) 252-125
7 (383-2) 356-005
Fax: 7 (385-2) 240-396
E-mail: iwep@iwep.altai.su
RUSSIA

Dr. Semen T. VASKOV
Institute of Automation and
Electrometry of SB RAS
Director
1 Universitetsky Ave.
630090 Novosibirsk
Tel: 7 (383-2) 354-550
Fax: 7 (383-2) 354-851
Telex: 133182 KWANT SU
E-mail: vaskov@iaie-phys.nsk.su
RUSSIA

Ms. Lynn WILKINSON
Center for Canadian-Russian Initiatives
President
6395 Elliott Str., 4B
Vancouver, BC V5S 2M5
Tel: 1 (604) 451-4561
CANADA

Dr. Edgar H. WINGENDER
GBF - Department of Genetics
Project Leader
Mascheroder Weg 1
D-38124 Braunschweig
Tel: 49 (531) 618-1203
Fax: 49 (531) 618-1202
E-mail: ewi@venus.gbf-braunschweig.dbp.de
GERMANY

Dr. Zurab A. YAKOBASHVILI
Deputy Minister of Science and
Technical Policy of the Russian Federation
11 Tverskaya Str.
103905 Moscow
Tel: 7 (095) 229-4774
Fax: 7 (095) 230-2823
RUSSIA

Prof. Kirill I. ZAMARAEV
International Centre for Catalysts
Characterization and Testing
Boreskov Institute of Catalysis of SB RAS
Director
5 Lavrentiev Ave.
630090 Novosibirsk
Tel: 7 (383-2) 355-750
Fax: 7 (383-2) 355-756
Telex: 133122 SOVET SU
E-mail: kiz@catalysis.nsk.su
RUSSIA

Prof. Vladimir I. ZLOBIN
Institute of Epidemiology and
Microbiology of East-Siberian Branch
Academy of Medical Sciences
Director
3 Karl Marx Str.
664000 Irkutsk
Tel: 7 (395-2) 333-445
RUSSIA

LIST OF OBSERVERS

Dr. Gennady N. ANOSHIN
United Institute of Geology,
Geophysics and Mineralogy of SB RAS
Director of Analytical Centre
3 Universitetsky Ave.
630090 Novosibirsk
Tel: 7 (383-2) 357-928
RUSSIA

Dr. Albert M. BEIM
Institute of Ecotoxicology
Director
665914 Baikalsk, Irkutsk Region
RUSSIA

Dr. Alex S. BOGOMOLOV
Physical-Technical Instrument Making
Department of SB RAS
Head of Department
3 Institutskaya Str.
630090 Novosibirsk
Tel: 7 (383-2) 357-022
RUSSIA

Dr. Valentin V. FURIAEV
Sukachev Institute of Forest of SB RAS
Senior Researcher
660036 Krasnoyarsk, Academgorodok
Tel: 7 (391-2) 459-660
RUSSIA

Dr. Illarion N. GLAZKOV
Institute of Semiconductor Physics of
SB RAS
Scientific Secretary
13 Lavrentiev Ave.
630090 Novosibirsk
Tel: 7 (383-2) 350-563
630090 Novosibirsk
RUSSIA

Dr. Alexandr E. KEL
Institute of Cytology and Genetics of SB
RAS
Senior Researcher
10 Lavrentiev Ave.
630090 Novosibirsk
Tel: 7 (383-2) 353-335
RUSSIA

Dr. Gennady G. KODENEV
Engineering Centre for Geophysical and
Ecological Instrument-Making of SB RAS
Deputy Director
3 Universitetsky Ave.
630090 Novosibirsk
Tel: 7 (383-2) 329-836
RUSSIA

Dr. Victor V. KOZLOV
Institute of Theoretical and
Applied Mechanics of SB RAS
Head of Laboratory
4/1 Institutskaya Str.
630090 Novosibirsk
Tel: 7 (383-2) 354-278
RUSSIA

Dr. Yury S. KUSNER
Institute of Geochemistry of SB RAS
Head of Laboratory
1a Favorsky Str.
664033 Irkutsk
Tel: 7 (395-2) 460-500, 460-220
RUSSIA

Dr. Victor I. KUZIN
Computing Centre of SB RAS
Head of Department
6 Lavrentiev Ave.
630090 Novosibirsk
Tel: 7 (383-2) 351-150
RUSSIA

Dr. Sergey V. KUZNETSOV
Siberian International Centre of
New Information Technologies
Senior Research Worker
6 Lavrentiev Ave.
630090 Novosibirsk
Tel: 7 (383-2) 350-579
RUSSIA

Dr. Stanislav S. KUROSOV
Institute of Carbon Materials
Chemistry of SB RAS
Senior Researcher
18 Sovetsky Ave.
650099 Kemerovo
Tel: 7 (384-2) 265-692
RUSSIA

Dr. Oleg A. MAKHOTKIN
Computing Centre of SB RAS
Chief Researcher
6 Lavrentiev Ave.
630090 Novosibirsk
Tel: 7 (383-2) 356-721
RUSSIA

Dr. Valery K. MALINOVSKY
Institute of Automation and
Electrometry of SB RAS
Deputy Director
1 Universitetsky Ave.
630090 Novosibirsk
Tel: 7 (383-2) 351-052
Fax: 7 (383-2) 354-851
E-mail: root@iaie-phys.nsk.su
RUSSIA

Dr. Leonid I. MALTSEV
Institute of Thermal Physics of SB RAS
Head of Department
1 Lavrentiev Ave.
630090 Novosibirsk
Tel: 7 (383-2) 357-402
RUSSIA

Prof. Boris P. MIRONOV
Institute of Thermal Physics of SB RAS
Head of Laboratory
1 Lavrentiev Ave.
630090 Novosibirsk
Tel: 7 (383-2) 357-342
RUSSIA

Dr. Vadim M. MORALEV
Institute of Inorganic Chemistry of SB RAS
Head of Laboratory
3 Lavrentiev Ave.
630090 Novosibirsk
Tel: 7 (383-2) 354-315
RUSSIA

Dr. Sergey V. MOROZOV
Novosibirsk Institute of
Organic Chemistry of SB RAS
Senior Researcher
9 Lavrentiev Ave.
630090 Novosibirsk
Tel: 7 (383-2) 351-662
RUSSIA

Dr. Sergey A. NAZARENKO
Tomsk Scientific Centre of RAMS
Institute of Medical Genetics
Head of Laboratory
634058 Tomsk, Sputnik
Tel: 7 (382-2) 772-823
RUSSIA

Dr. Valery A. ORLOV
Institute of Laser Physics of SB RAS
Head of Laboratory
13 Lavrentiev Ave.
630090 Novosibirsk
Tel: 7 (383-2) 350-947
RUSSIA

Dr. Nikolay M. OSCORBIN
Altai State University
Institute of Ecological Monitoring
Senior Researcher
66 Dimitrov Str.
656099 Barnaul
Tel: 7 (385-2) 251-023
RUSSIA

Dr. Victor B. SALYUKOV
Institute of Medical Genetics of RAMS
Senior Researcher
634058 Tomsk, Sputnik
Tel: 7 (382-2) 772-823
RUSSIA

Prof. Veniamin A. SIDOROV
Budker Institute of
Nuclear Physics of SB RAS
Deputy Director
11 Lavrentiev Ave.
630090 Novosibirsk
Tel: 7 (383-2) 356-031
Fax: 7 (383-2) 352-163
RUSSIA

Dr. Vladimir F. STARICHENKO
Novosibirsk Institute of
Organic Chemistry of SB RAS
Deputy Director
9 Lavrentiev Ave.
630090 Novosibirsk
Tel: 7 (383-2) 355-861
RUSSIA

Dr. Vadim A. STEPANOV
Institute of Medical Genetics
of Tomsk Scientific Centre of RAMS
Senior Researcher
634058 Tomsk, Sputnik
Tel: 7 (382-2) 772-823
RUSSIA

Dr. Jury A. VEDERNIKOV
Institute of Theoretical and
Applied Mechanics of SB RAS
Senior Researcher
4/1 Institutskaya Str.
630090 Novosibirsk
Tel: 7 (383-2) 356-920
RUSSIA

Dr. Boris V. YUDIN
Siberian Scientific Agency
Deputy General Director
23 Musa Dzhalil Str.
630055 Novosibirsk
Tel: 7 (383-2) 328-935
Fax: 7 (383-2) 328-932
RUSSIA

Dr. Alexandr A. ZEMLYANOV
Institute of Atmospheric Optics of SB RAS
Deputy Director
1 Academichesky Ave.
634055 Tomsk
Tel: 7 (382-2) 258-034
Fax: 7 (382-2) 259-086
Telex: 612510 LIDAR SU
E-mail: root@iao.tomsk.su
RUSSIA

LIST OF CONTRIBUTORS

Prof. Anatoly S. ALEKSEEV
Siberian International Centre of
New Information Technologies
Computing Centre of SB RAS
Director
6 Lavrentiev Ave.
630090 Novosibirsk
Tel: 7 (383-2) 355-650
Fax: 7 (383-2) 354-269
E-mail: aleks@comcen.nsk.su
RUSSIA

Dr. Marcel ANDRE
Centre d'Etudes Nucléaires de Cadarache
Conseiller Scientifique
13108 Saint-Paul-Lez-Durance Cedex
Tel: 33 (42) 257-578
Fax: 33 (42) 254-656
E-mail: thiery@orstom.orstom.fr
FRANCE

Prof. J.H. BLOCK
Fritz-Haber-Institut
der Max-Planck-Gesellschaft
Director
Faradayweg 4-6, Dahlem
D-14195 Berlin
Tel: 49 (30) 830-5411 or 830-5430
Fax: 49 (30) 830-5420
Telex: 185676 FHIMP D
GERMANY

Prof. Vladimir V. BOLDYREV
Institute of Solid State Chemistry of SB RAS
Director
18 Derzhavin Str.
630091 Novosibirsk
Tel: 7 (383-2) 200-423
Fax: 7 (383-2) 200-969
E-mail: boldyrev@solid.nsk.su
RUSSIA

Dr. Nikolay N. DYGALO
Institute of Cytology and Genetics of SB RAS
Chief Researcher
10 Lavrentiev Ave.
630090 Novosibirsk
Tel: 7 (383-2) 351-250
Fax: 7 (383-2) 356-558
RUSSIA

Dr. Boris S. ELEPOV
State Public Scientific Library of SB RAS
Director
15 Voshod Str.
630200 Novosibirsk
Tel: 7 (383-2) 661-860
Fax: 7 (383-2) 660-308
Telex: 133220 LIBRO SU
E-mail: elepov@libr.nsk.su
RUSSIA

Dr. Valery D. ERMIKOV
Presidium of SB RAS
Chief of Department
17 Lavrentiev Ave.
630090 Novosibirsk
Tel: 7 (383-2) 356-619
Fax: 7 (383-2) 354-846
E-mail: vde@hq.soan.nsk.su
RUSSIA

Prof. Yuri L. ERSHOV
Novosibirsk State University
Past Rector
2 Pirogov Str.
630090 Novosibirsk
Tel: 7 (383-2) 357-808
Fax: 7 (383-2) 356-237
E-mail: erchov@mioo.cnit.nsk.su
RUSSIA

Prof. Vasily M. FOMIN
Siberian Centre of Aerophysical Studies
Institute of Theoretical and
Applied Mechanics of SB RAS
Director
4/1 Institutskaya Str.
630090 Novosibirsk
Tel: 7 (383-2) 353-533
Fax: 7 (383-2) 352-268
Telex: 133128 POISK SU
E-mail: fomin@icas.nsk.su
RUSSIA

Dr. Valentin V. FURIAEV
Sukachev Institute of Forest of SB RAS
Senior Researcher
660036 Krasnoyarsk, Academgorodok
Tel: 7 (391-2) 459-660
RUSSIA

Prof. Josef I. GITELSON
International Centre for
Closed Ecological Systems
Institute of Biophysics of SB RAS
Director
660036 Krasnoyarsk, Academgorodok
Tel: 7 (391-2) 434-623
Fax: 7 (391-2) 222-556
Telex: 288139 MOST SU
E-mail: jgit@ibp.krasnoyarsk.su
RUSSIA

Dr. Johann G. GOLDAMMER
Max-Planck Institut for Chemistry
Leader of the Fire Ecology Research Group
D-79085 Freiburg
Tel: 49 (761) 808-011
Fax: 49 (761) 808-012
Telex: 4187674 MPCH D
E-mail: jggold@sun1.ruf.uni-freiburg.de
GERMANY

Dr. Bair GOMBOEV
Baikal Institute of
Nature Management of SB RAS
Senior Researcher
8 Sahjanov Str.
670042 Ulan-Ude
Tel: 7 (301-22) 38-035
E-mail: root@bien.buriatia.su
RUSSIA

Prof. Michael A. GRACHEV
Baikal International Centre for
Ecological Research
Limnological Institute of SB RAS
Managing Director
3 Ulan-Batorskaya Str.
664033 Irkutsk
Tel: 7 (395-2) 460-504
Fax: 7 (095) 420-2106 (Moscow)
E-mail: grachev@lin.irkutsk.su
RUSSIA

Dr. Paul P. HEARN
Office of International Geology
U.S. Geological Survey
National Centre MS 957
Deputy Chief
Reston, Virginia 22092
Telex: 248418 GEOINT UR
USA

Prof. Aleksandr S. ISAEV
International Forest Institute
Director
69 Novotcheremushkinskaja Str.
117418 Moscow
Tel: 7 (095) 332-8652
Fax: 7 (095) 332-6020
RUSSIA

Dr. Alain H. JUBIER
NATO, Scientific Affairs Division
Program Director
B-1110 Brussels
Tel: 32 (2) 728-4228
Fax: 32 (2) 728-4232
BELGIUM

Dr. Takayoshi KAWAI
National Institute for the
Environmental Studies
Global Environment Division
Senior Research Scientist
16-2 Onogama Tsukuba
Ibaraki 305
Tel: 81 (298) 516-111 Ext. 288
Fax: 81 (298) 514-732
E-mail: tkawai@npc60o.nies.go.jp
JAPAN

Prof. Anatoly M. KHARITONOV
International Center of Aerophysical
Studies
Executive Director
4/1 Institutskaya Str.
630090 Novosibirsk
Tel: 7 (383-2) 350-778
Fax: 7 (383-2) 352-268
E-mail: khar@icas.nsk.su
RUSSIA

Prof. J. KLERKX
Royal Museum of Central Africa
Head of Department of Geology
13 Leuvensesteenweg
B-3080 Tervuren
Tel: 32 (2) 769-5426
Fax: 32 (2) 769-5432
Telex: 27048 BUGEO B
E-mail: geomrac@vub.ac.be
BELGIUM

Dr. Nikolay A. KOLCHANOV
Institute of Cytology and Genetics of
SB RAS
Deputy director
10 Lavrentiev Ave.
630090 Novosibirsk
Tel: 7 (383-2) 353-335
Fax: 7 (383-2) 356-558
E-mail: root@cgi.nsk.su
RUSSIA

Prof. Valentin A. KOPTYUG
Siberian Branch of RAS
Chairman
17 Lavrentiev Ave.
630090 Novosibirsk
Tel: 7 (383-2) 350-567
Fax: 7 (383-2) 354-846
7 (095) 938-2097
Telex: 133128 POISK SU
E-mail: vak@hg.soan.nsk.su
RUSSIA

Dr. Gennady N. KULIPANOV
Siberian Synchrotron Radiation Centre
Budker Institute of Nuclear Physics of SB
RAS
Deputy Director
11 Lavrentiev Ave.
630090 Novosibirsk
Tel: 7 (383-2) 359-498
Fax: 7 (383-2) 352-163
Telex: 133819 KORUND SU
E-mail: kulipanov@inp.nsk.su
RUSSIA

Prof. Michael I. KUZMIN
Vinogradov Institute
of Geochemistry of SB RAS
Director
1a Favorsky Str.
664033 Irkutsk
Tel: 7 (395-2) 460-500
Fax: 7 (395-2) 462-952
E-mail: mikuzmin@igc.irkutsk.su
RUSSIA

Prof. Fedor A. KUZNETSOV
Institute of Inorganic Chemistry of SB RAS
Director
3 Lavrentiev Ave.
630090 Novosibirsk
Tel: 7 (383-2) 355-950
Fax: 7 (383-2) 355-960
Telex: 133226 VODA SU
E-mail: fk@che.nsk.su
RUSSIA

Prof. Günter LAUKIEN
Bruker Medizintechnic
President
Silberstreifen
D-7512 Rheinstetten 4 / Karlsruhe
GERMANY

Dr. Kirill G. LEVI
International Centre for Active Tectonics and Natural Disasters
Institute of the Earth's Crust of SB RAS
Deputy Director
128 Lermontov Str.
664033 Irkutsk
Tel: 7 (395-2) 461-562
E-mail: levi@cora.irkutsk.su
RUSSIA

Prof. Nikolay A. LOGACHEV
International Centre for Active Tectonics and
Natural Disasters
Institute of the Earth's Crust of SB RAS
Director
128 Lermontov Str.
664033 Irkutsk
Tel: 7 (395-2) 464-000
E-mail: log@cora.irkutsk.su
RUSSIA

Prof. Hermanus J. LOVINK
Akzo Catalysts
Senior Technical Associate
Stationsplein 4, P.O. Box 247
3800 AE Amersfoort
Tel: 31 (033) 676-317
Fax: 31 (033) 676-154
Telex: 79276 ACAF NL
The NETHERLANDS

Prof. Nikolay Z. LYAKHOV
Institute of Solid State Chemistry of SB RAS
Deputy Director
18 Derzhavin Str.
630091 Novosibirsk
Tel: 7 (383-2) 200-949
Fax: 7 (383-2) 200-969
E-mail: boldyrev@solid.nsk.su
RUSSIA

Dr. Jean-Pierre MASSUE
Council of Europe
European Federation of Scientific Networks
Secretary General
B.P. 431/R6, F-67006 Strasbourg Cedex 2
Tel: 33 (88) 412-614, 412-615
Fax: 33 (88) 412-787
FRANCE

Prof. Dr. Gerd E.A. MEIER
DLR-Institute for Fluid Mechanics
Director
Bunsenstrasse 10, 37073 Göttingen
Tel: 49 (551) 709-2177
Fax: 49 (551) 709-2889
Telex: 96839 DFVGOE D
E-mail: sigfried@spock.es.go.dkr.de
GERMANY

Prof. Vyacheslav I. MOLODIN
Altai International Centre for
Humanitarian and Biospheric Research
Institute of Archaeology and
Ethnography of SB RAS
Deputy Director
17 Lavrentiev Ave.
630090 Novosibirsk
Tel: 7 (383-2) 357-241
Fax: 7 (383-2) 357-791
RUSSIA

Dr. Sergey V. MOROZOV
Novosibirsk Institute of
Organic Chemistry of SB RAS
Senior Researcher
9 Lavrentiev Ave.
630090 Novosibirsk
Tel: 7 (383-2) 351-662
RUSSIA

Prof. Sten NILSSON
IIASA
Project Leader
Schlossplatz 1, Laxenburg
Tel: 43 (223) 671-521 Ext. 229
Fax: 43 (223) 671-313
Telex: 079137 IIASA A
E-mail: nilsson@iiasa.ac.at
AUSTRIA

Dr. Aleksey V. PODOPLELOV
Siberian Tomography Centre
Deputy Director
3 Institutskaya Str.
630090 Novosibirsk
Tel: 7 (383-2) 353-550
Fax: 7 (383-2) 352-366
Telex: 133148 KING SU
E-mail: root@tomo.nsk.su
RUSSIA

Dr. Vadim E. RYGALOV
International Centre for
Closed Ecological Systems
Executive Director
660036 Krasnoyarsk, Academgorodok
Tel: 7 (391-2) 434-623
Fax: 7 (391-2) 222-556
Telex: 288144 KIF SU
E-mail: abrek@ibp.krasnoyarsk.su
RUSSIA

Prof. Renad Z. SAGDEEV
Siberian Tomography Centre
Director
3 Institutskaya Str.
630090 Novosibirsk
Tel: 7 (383-2) 352-366
Fax: 7 (383-2) 352-366
Telex: 133148 KING SU
E-mail: root@tomo.nsk.su
RUSSIA

Dr. Boris G. SALTYKOV
Minister of Science and Technical Policy of
the Russian Federation
11 Tverskaya Str.
103905 Moscow
Tel: 7 (095) 229-2501
Moskow
RUSSIA

Prof. Suzanne A. SAVEY
Université Paul Valéry
Géographie et Aménagement du Territoire
Route de Mende, B.P. 5043
34032 Montpellier Cedex 1
Tel: 33 (67) 142-120 or 272-980
Fax: 33 (67) 142-332
FRANCE

Dr. Vyacheslav E. SELIVERSTOV
Siberian International Centre for
Regional Studies
Executive Director
17 Lavrentiev Ave.
630090 Novosibirsk
Tel: 7 (383-2) 353-954
Fax: 7 (383-2) 355-580
E-mail: sel@sicrs.nsk.su
RUSSIA

Prof. Vladimir K. SHUMNY
Altai International Centre for
Humanitarian and Biospheric Research
Scientific Leader of Biospheric Research
Institute of Cytology and Genetics of
SB RAS
Director
10 Lavrentiev Ave.
630090 Novosibirsk
Tel: 7 (383-2) 354-350
Fax: 7 (383-2) 356-558
E-mail: root@cgi.nsk.su
RUSSIA

Dr. Gennady Ya. SMOLKOV
Institute of Solar-Terrestrial
Physics of SB RAS
Deputy Director
664033 Irkutsk, Box 4026
Tel: 7 (395-2) 463-791
Fax: 7 (395-2) 462-557
E-mail: root@icc.ccsoan.irkutsk.su
RUSSIA

Dr. Arnold K. TULOKHONOV
International Scientific Centre of Socio-
Ecological Problems of the Lake Baikal
Basin
Baikal Institute of Nature Management of
SB RAS
Director
8 Sahyanova Str.
670042 Ulan-Ude
Tel: 7 (301-22) 38-035
Fax: 7 (301-22) 63-444
Telex: 219117 LOTOS SU
E-mail: root@bien.buriatia.su
RUSSIA

Dr. Evgeny A. VAGANOV
Siberian International Centre for
Ecological Research of Boreal Forests
Sukhachev Institute of Forest of SB RAS
Director
660036 Krasnoyarsk, Academgorodok
Tel: 7 (391-2) 494-695
Telex: 288018 SOSNA SU
E-mail: dndr@ifor.krasnoyarsk.su
RUSSIA

Prof. Francis Van NOTEN
Royal Museums of Art and History
Director
10 Jubelpark
B-1040 Brussels
Tel: 32 (2) 741-7200
Fax: 32 (2) 733-7735
BELGIUM

Prof. Douglas F. WILLIAMS
University of South Carolina
Department of Geological Sciences
Project Leader
Columbia, SC 29208
Tel: 1 (803) 777-7525
Fax: 1 (803) 777-6610
E-mail: doug@epoch.geol.scarolina.edu
USA

Prof. Kirill I. ZAMARAEV
International Centre for Catalysts
Characterization and Testing
Boreskov Institute of Catalysis of SB RAS
Director
5 Lavrentiev Ave.
630090 Novosibirsk
Tel: 7 (383-2) 355-750
Fax: 7 (383-2) 355-756
Telex: 133122 SOVET SU
E-mail: kiz@catalysis.nsk.su
RUSSIA

Prof. Gely A. ZHEREBTSOV
Siberian Centre of Solar-Terrestrial Physics
Irkutsk Scientific Centre of SB RAS
Chairman of Presidium
664033 Irkutsk, Box 1249
Tel: 7 (395-2) 460-565
Telex: 133163 TAIGA SU
E-mail: root@sitmis.irkutsk.su
RUSSIA

I. CURRENT PROBLEMS AND PROSPECTS OF INTERNATIONAL COLLABORATION

NEW RUSSIAN PROBLEMS OF INTERNATIONAL SCIENTIFIC COLLABORATION

B.G. SALTYKOV
Minister of Science and Technical Policy of the Russian Federation, Moscow, Russia

ABSTRACT. Russia's formal independence required the formation of new principles in science policy, including deideologization, openness of society and pragmatism in international scientific collaboration. The orientation toward contacts with researchers of the European Community and search for new forms of collaboration, such as INTAS and NATO grants, and foundation of international scientific centres in Russia - these are the signs of today. Deficiency in the Russian legislation in science and technology, customs regulation, absence of contract law, property problems are dealt with at the Ministry of Science and Technical Policy at the present time.

Keywords: science policy / scientific collaboration / new principles / new forms.

The workshop "Scientific Policy: New Mechanisms for Scientific Collaboration between East and West" envisages a discussion of concrete scientific problems. Its participants are working, I would say, at the microlevel of international scientific co-operation. By virtue of my duties, I am going to dwell on macro-problems the Russian government and the Russian science face at this new difficult stage of development.

The disintegration of the USSR at the end of 1991 and acquisition of formal sovereignty by Russia have required re-conclusion and re-consideration of many agreements and treaties with foreign countries. Generally speaking, they could be just prolonged, since Russia has declared herself, and has been recognized as the successor of the USSR. However, in the field of international scientific and technical collaboration many aspects have appeared which do not permit us to automatically follow the obligations of the USSR. I will focus briefly on three new aspects of our collaboration: political, economic, and properly scientific ones.

Still as early as the beginning of 1992, during negotiations with foreign scientific and technical organizations and with governments of various countries, we emphasized the new political aspects of collaboration, the principle of which consists in the fact that Russia has renounced the autarchic, closed mode of existence. Of course, in the late 80s and early 90s, the USSR was already making steps toward this, but now we have declared this as the principle of our policy - openness to the scientific community and a pragmatic approach to international scientific collaboration.

This means also that we renounced the ideologization which was present in our scientific collaboration (unfortunately, on a large scale), e.g. in setting up large

V A Koptyug and J Klerkx (eds), Science Policy New Mechanisms for Scientific Collaboration between East and West, 3–8

international centers. I cannot say it dominated, but even the creation of the Joint Institute of Nuclear Research in Dubna pursued also political goals. Luckily, the collaboration and relationships between the scientists themselves corrected to a high degree this policy of a certain part of Soviet authorities, and it is only thanks to the scientists that Dubna at present is a real international center professing the principles of a civilized scientific community. During the last two years, we succeeded in eliminating an atmosphere of mutual suspicion. Our present task is to consolidate the atmosphere of mutual trust which is being formed and has largely been formed not only in the relations between scientific communities of different countries, but also between persons responsible for this collaboration.

In order that the collaboration develops successfully further, in order to establish long-term normal relations between scientists, the political situation within Russia must be stable.

Economic aspects may be a topic of a separate discourse: today they permeate all our talks, problems and relationships. To be brief, Russia is going through a deep economic crisis. This means that there is an enormous budget deficit, that there is no (and cannot be under the conditions of recession in production) demand for innovations and for new technologies, and that the scientific and technical sphere in Russia is in an extremely hard situation.

During the last 2—2.5 years, quite a new phenomenon has appeared: the international scientific community has members in government, and a whole infrastructure and movement have arisen in support of the Russian fundamental science which is virtually an international property. This has become possible also as a result of a thorough study of the state of science in Russia, the so-called "case study" on the spot, and as a result of a long history of relations between scientists of different countries, which have become the foundation for mutual trust. The previously existing form of bilateral collaboration: the so called moneyless exchange today is rarely employed, for in this type of arrangement the expenses of the visiting scientist must be covered in full by our side. This is quite difficult.

The numerous funds that have appeared make it possible to pay for a considerable part of bilateral collaboration through foreign sources of financing. The most typical example of the last few months is the creation of a European association for the support of fundamental science in the countries of the former USSR — the product of nearly two-years work of the international scientific community — both European and Russian. There is pleasant news: the Council of this association has decided to spend in one year 21 million ecus in European funds, that had been previously earmarked for a two year period. I will remind you that a 2 + 1 scheme (2 collectives on the part of the European Community, and 1 on the part of the FSU countries) is accepted there, and the history has disposed so that Russia gets on the average 75—80 % of this support, according to her scientific potential (although this percentage varies for different fields of science).

A new aspect has become evident with the possibility of starting direct collaboration between Russian and foreign scientific teams. This fact speaks for itself, because in the past, quite often there was collaboration not with the scientists themselves but with bureaucrats from the scientific community. On the other hand there was scientific tourism that was characterized by trips aboard of certain groups, often scientists. But of course, later these groups vanished without a trace, and collaboration in this respect came to a halt.

A characteristic example of ideologization of collaboration was, unfortunately, the so-called comprehensive program of scientific and technical development of Comecon countries. There was a sensible aspect to this long term program. There, many teams that found worthwhile partners, however the scope and forms of collaboration witnessed did not possess a scientific nature.

Unfortunately, at present, we have been renouncing our collaboration with the countries of Central Europe too vigorously. The fact is that when more remote Western countries became accessible to collaboration, a certain euphoria ensued in our scientific circles, and they stopped paying attention to the countries of the Near West or, shall we say, of the middle foreign countries. A certain role was played in this by our new pragmatic approach to the international scientific collaboration. While formerly the trips to Comecon countries were almost free of charge and were just a kind of "running-in", now everything must be paid in hard currency even there, so that Russians will go there only if there are really some results and mutual interest. That is why, a rearrangement of the collaboration structure is taking place. For the last 2 years it has been evident that whole trends of exchange and trips, formerly inflated, are dying out. On the other hand, new contacts are arising. Recently, Russia signed a new agreement for scientific and technical collaboration with Hungary. Our common opinion is that it is time to "gather stones" scattered on the field of former collaboration and to continue joint research which is useful to the scientific collectives of the two countries.

A special problem is imperfection of Russian legislature in the sphere of science and technology. It lacks answers to most questions posed by everyday life. For example, we still have no definite concept of a charity fund, no full-value charity law, no law of commercial non-profit organizations etc. We do not have these things because the structure of relationships had been different.

Something has been done by joint efforts of the Ministry of Finances, tax inspection, and committees of the former Supreme Soviet. At present, grants given by charity organizations are not taxed. However if a foreign university or worse, a firm gives a grant, the grantee is obligated to prove that the university or firm is not pursuing commercial interests and that that which is being given is a grant and not a salary by contract. This is necessary because the tax service is concerned that certain high paying commercial activities in the techno-scientific field will be concealed (without taxation!) as grants.

Although, it seems that in the field of fundamental research, where gaining a patent, profit making, etc. is not an issue, and where collaboration is conducted on a more spiritual basic, the problem basically boils down to organizing and coordinating such researches and their financing. During the last year or year and a half, about 5 or 6 international funds have been set up; the most well-known is the Soros Foundation (it has gone as far as Siberia). I am not going to talk about the problems that were arising in the process of its setting up and organizing the possibilities for its work — these are our internal problems. But the question that concerns all the scientists: namely the abolition of grant taxation — so far has not been resolved legislatively.

Another painful problem is that of questions associated with customs (value-added tax, customs tariff, and payment for demurrage and storage). We reach a paradox when we are forced, for example to pay for literature that comes in the form of a grant. While we spend a month struggling with customs to prove that the received literature is a grant and therefore not subject to tariffs, customs will bill us for the storage of the literature. At present the material influx including charity, across the boarder is unregulated. A special problem is the temporary export or the one, two or three year import of scientific equipment for conducting experiments. Here again we need to regulate in a reasonable fashion the amount of customs tariffs.

There is a complete absence at present, of any written law especially in the techno-scientific field: this was also a case in the former USSR. Our scientists, designers and engineers are quite naive in this respect: they do not know either the procedure or their own rights, or the world prices. There have been cases when concrete results of research were sold for outrageously low prices.

This is an omission on the part of the government and its structures which have to teach, explain, and create laws. Therefore, even I take some of the blame. At the same time, I am appealing to my western colleagues: who could render great methodological help to us by explaining the foundations of contract law and types of relationships in the techno-scientific field. I will focus one type of scientific collaboration: the creation of foreign science centers on Russian territory. In principle this is a kind of barrier to "brain drain" or the outward flow of our emigrant scientists who as a matter of fact export their expertise directly to the place where it is used. This from collaboration has at least too advantages: for us the advantage is that our people remain here and get an opportunity to work with the state-of-the-art computer and laboratory equipment: for the foreign partner, the advantage is that comparable results can be obtained at a lower cost for the firm. Some recent examples of this can be seen in Moscow where an affiliate of Boeing that will collaborate with the Central Institute of Aero- and Hydrodynamics recently opened. Soon, an affiliate of the Samsung Company will open.

One of the most difficult problems here is that of distribution of intellectual property. It has not been solved so far. We are against any exclusive rights of foreign investors, and often we have mutual understanding, but in practice it is very difficult to control the export of intellectual products, of know-how.

At present, the Russian State has loosened considerably the control of export of technologies, which can now be exercised either when documentation and specimens are exported, or when a person possessing know-how goes abroad.

Of course, we are not going to change our principles by limiting freedom of entry and departure, but a document defining the procedure of export of technologies is in preparation. It is not an easy matter. On the one hand, we risk of returning to the past by establishing rigid bureaucratic controls, where the decision rests almost completely in the hands of bureaucrats. This can completely block all developing contracts. On the other hand one can understand the concern of not only the security agencies, but of the institutes which possess intellectual potential: since private individuals who do not have obligations to the state or to an institution can export that that may have taken a scientific team, not the individuals, decades to create.

Another difficulty is the absence of any fixed rules of existence and activity of international scientific centers in the territory of Russia. The mentioned Joint Institute of Nuclear Research in Dubna which has existed since 1957 is today in a complete juridical vacuum; meanwhile, there is a lot of controversial questions (taxes, land property, pension fund etc.). Unless regulated by legal agreements, they will always complicate the activities of the international center in the Russian territory.

Still another typical circumstance is the lack of infrastructure in the centers of collaboration. Russia is known to be characterized by an enormous deficit of infrastructure — telecommunications, hotels etc. Under the present economical conditions, those who have created this infrastructure are in a better position. For example, the Siberian Academgorodok, although it has its difficulties, has nevertheless a great advantage for continuing international collaboration.

The last aspect is mutual relationship between Russia and the countries of the former Soviet Union. This is also a problem of international collaboration. In the first half of 1992, this looked outrageous, we could not overcome our mentality. However, at that time, unfortunately, a wild expropriation of former Russian property was going in the territory of those countries, and after two years, this problem is still the principal one: the questions of division of rights of property and definition of using these rights in the territory of these countries have not been settled. The most known examples: in Ukraine, the radiotelescope situated in Crimea, in Uzbekistan — another system of telescopes. It is becoming gradually clear that the sovereign republics cannot use these installations without the help of Russian scientific centers, they cannot complete their construction or even maintain them. We are now working on intergovernmental agreements intended to protect the rights of scientists of the two countries (in cases of such installations) or of all the countries. One of ways out is setting up international centers and laboratories, e.g. for the astronomy complex in Uzbekistan. But again, investments are necessary for this.

Recently, we signed a new agreement with CERN — a well-know European company. Why was it not possible to continue the collaboration in the framework of the former USSR—CERN agreement? Partly because formerly Kazakhstan, Ukraine and Armenia participated in this agreement as one unit, and mutual commitments, e.g. to supply transformers from Zaporozhye or conductors from Kazakhstan were taken on behalf of Moscow. To-day this is not the case. Today, Ukraine and Kazakhstan raise the question of entering into CERN, although it seems to be a matter of the distant future. We have to define clearly the possibilities of Russian institutes, Russian science, and act within the limits in which we can be responsible for the results of the collaboration.

Concerning the role of the State in the international scientific collaboration, we have enunciated a principle of handing all the collaboration down to the level of the institutes, teams and laboratories which will directly participate in it. It is becoming clear that this is efficient and good for starting or continuing the contacts, although sometimes the partners demand guarantee from the State, not only financial, but also juridical one. This is mainly not the case with fundamental, but with applied researches.

Russia has recently practically entered into the European "EUREKA" (at present, a governmental decree of formal entry is being prepared) which regulates such relationships, especially those associated with transportation of high technologies across the frontiers.

However, nowadays, the habitual relationships between the State and the institute have changed radically; even the largest projects are sometimes made without any intervention of the State and technical complex — the Khrunichev plant in Moscow, a producer of rockets of "Proton" type — concludes contracts with foreign companies for commercial launching of satellites on its own, and today we often have difficulties with relations between such subjects of right and the State.

Sometimes such transactions at the level of institutes and design offices affect intergovernmental relations. For example, the progress in the relations between Russia and USA due to Russia's participation in the construction of the large space station "Freedom" (now, it will be called Alpha). But this immediately affected Russia's relations with Europe, the European Space Agency, and French agencies in the same field. We are now inventing, already at the level of government officials, dampers to quell these "waves" and explain our policy. I hope that in future years we will be able to consolidate what we have succeeded in achieving here, in Russia and Siberia, in this short time.

♦

NATO INTERNATIONAL SCIENTIFIC EXCHANGE PROGRAMS

A.H. JUBIER
NATO, Scientific Affairs Division, Brussels, Belgium

ABSTRACT. The NATO Science Program today involves the encouragement of collaboration between East and West. Various mechanisms of support of priority areas are developed to implement this collaboration.

Keywords: priority areas / funding of research / mechanisms of support.

It is a great pleasure for me, on behalf of NATO and its Science Committee, to thank you all for coming to this meeting in order to address issues of international scientific cooperation between East and West. I will make a brief presentation of the NATO Science Program, and I will certainly be glad to answer any questions, in particular specific questions about how to set up a cooperation and how to get support for it. Therefore, during the whole duration of the workshop, please feel free to come up to me and discuss any matter regarding support from the NATO Science Program.

In addition to its well-known political and military dimensions, NATO has a "Third Dimension" which seeks to encourage interaction between peoples, to consider some of the challenges facing our modern society and to foster development of Science and Technology. The programs of the NATO Science Committee for scientific cooperation are indeed a major component of this Third Dimension.

The NATO Science program was created as early as 1958, and until 1991 was focused on exchanges between scientists from the 16 countries of NATO(*), although not exclusively, with the enhancement of science and technology as its objective. Starting in 1992, due to the new geopolitical situation, a new era began for the science program which was reoriented in order to allow for international collaborations and exchanges of scientists and engineers between East and West. This new action, known as the Outreach Activity, is addressing some scientific and technological problems now being encouraged by 22 countries (the former USSR and most of Central Europe) which are now NATO's Cooperation Partner (CP) Countries(**).

(*) NATO member countries: Belgium, Canada, Denmark, France, Germany, Greece, Iceland, Italy, Luxembourg, Netherlands, Norway, Portugal, Spain, Turkey, United Kingdom, United States.

(**) Cooperation Partner (CP) countries: Albania, Armenia, Azerbaijan, Belarus, Bulgaria, Czech Republic, Estonia, Georgia, Hungary, Kazakhstan, Kyrgyzstan, Latvia, Lithuania, Moldova, Poland, Romania, Russia, Slovak Republic, Tajikistan, Ukraine, Uzbekistan.

V. A. Koptyug and J. Klerkx (eds.), Science Policy: New Mechanisms for Scientific Collaboration between East and West, 9–11.

Following consultation with those 22 countries, the following four priority areas were initially selected for particular support under the Outreach Activity:

Priority Area 1 — **Disarmament Technologies:** scientific problems related to disarmament technologies, including the disposal of nuclear and chemical weapons, defense industry conversion and the safety of nuclear technologies.

Priority Area 2 — **Environment:** scientific problems related to the environment, including the reclamation of contaminated military sites, regional and environmental problems (e.g. the Baltic Sea, the Black Sea, the Aral Sea and the Russian Arctic), and natural man-made disasters.

Priority Area 3 — **High Technology:** scientific problems related to high technology, including information science, material science, biotechnology, energy conservation and supply (non nuclear).

Priority Area 4 — **Human Resources/Science and Technology Policy:** problems related to human resources, including science policy, management, intellectual property rights and career mobility (e.g. the redeployment of defense industry scientists). This workshop that you are presently attending is indeed supported by this fourth priority area, having to do with international collaboration, exchange of scientists and science policy in general.

Finally, in order to deal with the above problems in the best possible conditions and enhance the scientific cooperation between East and West, we believe that the most advanced communication technologies, such as computer networking, should be used. Therefore, additional funding could be provided for computer networking expenditure once an award is made for scientific cooperation.

To implement East-West collaboration within the above priority areas, the following mechanisms for support are possible:

Advanced Research Workshops (ARWs):

The objective of an Advanced Research Workshop is to contribute to the critical assessment of existing knowledge on new important topics, to identify directions for future research, and to promote close working relations between scientists from different countries and with different professional experience. They are working meetings of between 20 and 50 participants lasting from two to five days. ARWs in priority areas should be co-directed by a scientist from a CP country and a scientist from a NATO country; they may be held in either a NATO country or a CP country. Funding is provided to ARW directors to cover organizational expenses, the travel and living expenses of key speakers and to contribute to the travel and living expenses of other selected participants. (N.B. The present workshop is typically co-directed by Professors Koptyug and Klerkx and held in Novosibirsk. However, the total number of participants is untypical and does exceed the average because of its special topics in science policy).

Advanced Study Institutes (ASIs):

The objective of an ASI is to promote the dissemination of scientific information, to impart experience and to foster professional contacts among scientists. Opportunities are offered to leading scientists (ASI directors) to organize advanced meetings of a tutorial character, typically of a duration of ten working days, involving 12 to 15 lecturers with an audience of 60 to 80 post-doctoral level "students". ASIs in the priority areas should be co-directed by a scientist from a CP country and a scientist from a NATO country; they may be held in a NATO country or a CP country. Funds are provided to ASI directors to cover organization expenditures, the travel and living expenses of lecturers and to contribute to the travel and living expenses of selected ASI students.

Collaborative Research Grants (CRGs):

The objective of a CRG is to stimulate collaboration between research scientists. CRGs are available for collaboration between researchers in NATO countries and researchers in Cooperation Partner countries in the priority areas. Funding is available for reciprocal visits abroad of from one to four weeks for one or two members of research teams collaborating on a joint research project. Awards are for one year, after which applications can be considered for a further year of funding.

Linkage Grant (LGs):

The objective of a LG is to promote cooperation between research teams from NATO countries and research teams from Cooperation Partner countries, with the aim of undertaking joint research projects that are in the priority areas. Funding is available for reciprocal short visits abroad for four or five members of each research team over a period of one to two years. Linkage grants may be bilateral or multilateral. For laboratories in Cooperation Partner countries a contribution may be requested towards acquisition of scientific equipment of modest cost essential for and specific to the project. Computer networking supplements are also available for CP country teams following the award of a Linkage Grant.

Expert Visits (EVs):

This programme supports the travel and living expenses of experts from NATO countries visiting research laboratories in Cooperation Partner countries for consultation on research projects in the priority areas.

In addition to these five mechanisms for support, the NATO Science Program is also providing Science Fellowships to enable scientists to continue their training or carry on research in a country other than their own for a duration of several months up to one year. These opportunities are offered in a limited number to scientists from Cooperation Partner (CP) countries for tenure in a NATO country, as well as for scientists from NATO countries for tenure in a CP country.

In summary, NATO support (1) can be made available for East-West meetings or joint projects provided their topics are within the priority areas, and (2) can only cover travel and living expenses, plus advanced communication expenditure.

Although the NATO Science Program is offering limited funding it must be kept in mind that, over the past 30 years, tens of thousands of scientists have benefited, including some from Eastern countries. Today is the time for expanding our support for East-West collaboration.

To conclude my presentation, I would like to wish you all a most successful meeting, with the hope that the NATO Science Program can help you set up new collaborations.

♦

A NETWORK OF RESEARCH CENTRES IN SIBERIA, AS A BASIS FOR MULTI-DISCIPLINARY INTERNATIONAL SCIENTIFIC COLLABORATION

V.A. KOPTYUG
Chairman of the Siberian Branch of the Russian Academy of Sciences, Novosibirsk, Russia.

ABSTRACT. Human civilization is at a turning point of history - the mankind must find a way out of the global crisis into the sustainable future. In the changing world the role of science will be rapidly growing, and the organization of scientific community will suffer changes in order to take up challenges of our time. As a result, some new tendencies are developing in science policy giving priorities to more goal-oriented, multidisciplinary international projects and programmes. In this context a possible role of the network of the International Research Centres in Siberia is discussed.

Keywords: sustainable development / science policy / international research centres / integration of science and education.

Recently we've been using the words "in transition" - countries are in transition, economical systems are in transition, science is in transition, and so on. What are the reasons for this reiteration and what are the goals of transition ? Transition from what to what?

When we speak about Russia, we have in mind first of all the crisis in the political and economic systems, and attempts to resolve it. When the NATO Science Committee discusses necessary changes in science policy, it first of all takes into account the recent trends in the changing geopolitical situation.

I believe that in all discussions of this kind one main factor should be stressed - the global crisis of our civilization, caused by a very rapid growth in population, by the uneven economic development of various parts of the world and by development with disregard to nature.

Mr. Maurice F. Strong, Secretary-General of the United Nations Conference on Environment and Development (Rio de Janeiro, June 1992) stressed in the Statement at the opening of the Conference: "The same processes of economic growth which have produced such unprecedented levels and power for the rich minority have also given rise to the risks and imbalances that threaten the future of rich and poor alike. This growth model, and the patterns of production and consumption which have accompanied it, is not sustainable for the rich, nor can it be replicated by the poor. To continue along this pathway could lead to the end of our civilization".

V. A. Koptyug and J. Klerkx (eds.), Science Policy: New Mechanisms for Scientific Collaboration between East and West, 13–17.

Therefore, we are now at a turning point in the history of human civilization. We must find a new path into the future. The abovementioned UN Conference on the Environment and Development (UNCED) indicated this road — the road of sustainable development, reconciling the problems of economic policy, development and the environment.

Therefore today it is becoming a necessity to consider all decisions taken on all levels — on the level of governments, on the level of the Presidium or General Assembly of the Russian Academy of Sciences (RAS), or on the level of the NATO Science Committee — through the prism of Rio's Agenda 21, that is, from the point of view of sustainable development requirements.

In relation to science the requirements may be found in Agenda-21 and in the materials of the International Conference "An Agenda of Science for Environment and Development into the 21st Century" organized by the International Council of Scientific Unions on the eve of UNCED in November 1991.

The role and responsibility of science in the society of the future will drastically increase, since the global problems of the environment and development require global decisions based on global conclusions made by the scientific community. The scale and complexity of the problems to be solved introduce some new requirements to the organization of scientific research and expertise. Among them I would like to emphasize the necessity of strengthening the system analysis, inter- and multidisciplinary approaches and the efficiency of coordination of scientific efforts on national and international levels. In my opinion this means that the organizing components in the infrastructure of science will increase.

In the majority of countries, financial resources available for scientific research are limited, and therefore the selection of priorities, and the subsequent concentration of national and international scientific potential and efforts on these priorities, are of very high importance.

In this situation the optimal balance between individualism and collectivism (both playing important roles in science and its organization) will be shifted additionally to the side of collectivism. But it is quite typical for the activity of foreign science foundations in Russia to use a mainly individual approach in supporting science. Such foundations pursue two goals: to support the most promising scientists on a benevolent basis and to obtain detailed information on such scientists for future purposes. In the current situation we are grateful for any activity. We appreciate the support to our individual scientists by, for example, a grant of five hundred dollars, since our government is unable or does not want to fulfill one of the recommendations of Agenda-21:

"Creating conditions (e.g. salaries, equipment, libraries) to ensure that the scientists will work effectively in their home countries" (item 35.25 a). But if Western sponsors wish to support science in Russia on a long-term basis it would be better to stimulate corresponding long-term collaboration in priority areas of basic and applied research, where we can play the role of equal partners. This corresponds to the recommendation of the United Nations — "to organize global partnerships in science for sustainable development".

No doubt, the fate of an individual scientist is very important, but the fate of high level scientific schools is no less important for world science.

In order to stimulate the involvement of scientific schools in broader international collaboration and to support the multidisciplinarity of such collaboration, the Presidium of the Siberian Branch (SB) of the RAS launched a special Programme two years ago on the Development of International Research Centres in Siberia. We have created 15 such centres and some of them are already efficiently acting as "open international laboratories" where scientists from various countries can work together with Russian scientists, using the sophisticated and diverse facilities of some institutes of SB RAS. We believe that these centres may be considered as elements of a distributed infrastructure network, supporting the integration of world science.

The joint research of the Siberian International Centres is usually concentrated on the most important natural objects of Siberia (such as Lake Baikal, Altai Mountains region, etc.) or on promising technical or technological subjects (synchrotron radiation, catalysts characterization and testing, etc.) A set of specific projects are approved each year by the International Scientific Council of each centre.

The duration of work by foreign scientists in the centres varies from 1 to 3 months a year (participation in expeditions or in experiments, using sophisticated equipment and installations), with the subsequent treatment of the data and results obtained usually being carried out in their home countries.

Our experience shows that on the basis of such centres some expanding teams of scientists from various countries can be gradually formed to collaborate regularly with their Russian colleagues. It seems that NATO's Collaborative Research Grants sub-programme is very suitable for attracting new participants to such scientific groups spread among many NATO countries. The objective of Collaborative Research Grants is "to stimulate collaboration between laboratories in different countries and thus enhance the effectiveness of research". This objective will be achieved in such cases.

Many papers from this Workshop are devoted to the activity of the abovementioned International Research Centres in Siberia. In this connection I would like to give some brief information about Siberia and the Siberian Branch of the Russian Academy of Sciences.

Siberia is a vast and very rich land that covers an area of 10 million square kilometres, has large reserves of minerals and other natural resources and is characterized by a wide geographical diversity.

Forty years ago the Government of the USSR developed a giant plan oriented to the accelerated development of Siberia and the Far East. The implementation of that plan required serious scientific support and, in order to fulfill this requirement, the Siberian Branch of the former USSR Academy of Sciences (now — the Russian Academy of Sciences) was established.

At present the SB RAS has Scientific Research Centres in Tyumen, Omsk, Novosibirsk, Tomsk, Kemerovo, Krasnoyarsk, Irkutsk, Ulan-Ude and Yakutsk, which altogether represents more than one hundred research institutes and development organizations.

Similar regional branches of the RAS were later organized for the Urals and the Far East regions.

The regional branches function according to the Charter of the RAS but receive financial support separately from the federal budget.

The achievements of Siberian scientists in basic research are well known in the world. However, I would like to underline that SB RAS, as a scientific community, carries out not only basic research, but is also involved to a great extent in the development of cleaner technologies for various branches of industry. In this respect we have, within the framework of SB RAS, a set of design organizations and pilot plants. This fact explains the close interaction of SB RAS with Siberian industry.

I emphasize this feature of our community in connection with the tendency relating to science policy in changing world, mentioned in a recent issue (№38) of the Quarterly Bulletin "NATO Science and Society Newsletter" in the article "Science and NATO in Transition".

In this paper Dr. Edward E. David, the United States representative on the NATO Science Committee, considers a trend relating to science policy or more precisely — to the research paradigm. During the last 45 years science and society in the developed countries professed a paradigm based on the affirmation that "investigator-initiated research is the bedrock of both knowledge and technology". Now, taking the global problems of our civilization into account, a tendency emerges to orient science toward "more goal-oriented research".

In my opinion some reorientation is inevitable. The role of goal-oriented research will certainly be increased but at the same time the new paradigm is very risky. There is a danger that in its framework governments and politicians will try to reorganize science according to their points of view.

Dr. Edward E. David believes, and I agree with him, that in relation to research, in this situation "an important safeguard is available — that is, the strict maintenance of quality". It was just this circumstance that we had in mind, when we made the decision to create in Siberia a set of International Research Centres.

There exists an additional attractive feature of the International Research Centres located in the Academic towns of SB RAS. A spirit of multidisciplinarity stimulates the establishment of broader international collaboration in the neighbouring basic and applied fields of research. This is very important, in particular, in studies of the complex scientific problems related to the implementation of the sustainable development conception.

The United Nations Organization is anxious about slow implementation of the Agenda-21 recommendations, due primarily to limited financial resources on national and international levels. In regards to this the Siberian Branch of the RAS proposes the selection of several specific regions in some countries of the world, the sustainable development of which is crucial not only for the particular country but also for the entire human race. These territories should be considered as "the world's model territories" to be periodically controlled by UN missions with regard to the fulfillment of a well balanced programme of economic, social and environmental development (sustainable development) corresponding to the recommendations of Agenda-21. The concentration of available financial resources on "model territories" will make results visible and therefore more comprehensible to the country's population.

In the course of a meeting of the High Level Advisory Board under the UN Secretary-General I mentioned the region of Lake Baikal as a possible example.

Lake Baikal is the most ancient (20—30 million years), the deepest (1632 m maximum depth), one of the largest (20% of the world's deposit of fresh water) and cleanest lakes on the planet. Baikal is inhabited by some 2000 species, 2/3 of which are endemic, i.e.they do not occur elsewhere on the Earth. The bottom sediments of Lake Baikal (thickness up to 6—7 km) contain the record of Central Asia paleoclimate over many millions of years.

The industry around Lake Baikal is mainly located in industrial districts gravitating towards the cities of Irkutsk, Angarsk, Shelekhov and Ulan-Ude. In other areas there are only some isolated plants (point sources of emissions). Therefore the territory of the Lake Baikal region is polluted only locally — by emissions from non-ferrous metallurgy plants, from the chemical industry and by the mining industry in the eastern part of the Baikal region.

The location of the pollution in this region allows the transport and transformation of pollutants to be studied separately for various branches of industry. The water in Lake Baikal is mainly polluted by the atmosphere and by the Selenga river running from Mongolia across Buryatia into the Lake.

There exists a national programme devoted to the protection of the Lake and to the economic development of its basin area. In the framework of this programme the industry of the region is strongly involved in collaboration with Western companies.

SB RAS, as I mentioned, is also involved in the development of cleaner technologies for many branches of industry. This circumstance, in combination with the transfer of advanced Western technologies, ensures favorable conditions for sustainable development of the region under consideration.

Thus, the Lake Baikal region is essential for mankind as the largest freshwater source, as a unique site of endemic biodiversity and as a "natural laboratory" for long-term international collaboration. It already has well developed international scientific liaisons and a good scientific background, and is characterized by the evident tendency of moving towards cleaner production. Therefore it seems possible to consider it as a candidate for the system of "the world's model territories" for sustainable development.

Returning to the subject of the International Research Centres, it seems desirable to draw attention to their usage in the context of the Advanced Study Institutes of the NATO International Scientific Exchange Programme, bearing in mind the objective of this Programme — "to disseminate advanced knowledge not yet in the university's curricula". The scientific institutes of SB RAS have good experience in collaborating with universities in the areas of education and training of young specialists.

In conclusion I would like to express my satisfaction with the first contacts of the Siberian Branch of the RAS with the NATO Science Committee and Scientific Affairs Division: they were very promising.

I hope that this Workshop and the discussions of new instruments and mechanisms of international collaboration between East and West will further stimulate this collaboration.

Speaking of the collaboration of the NATO Scientific Council with the Russian Academy of Sciences as a whole and with scientists from universities and industry, it seems that the creation of a special Scientific Council, by the Ministry of Science and Technical Policy of the Russian Federation for supporting and coordinating collaboration with NATO countries' organizations, will be very helpful in promoting large-scale long-term collaboration in the high priority areas.

References

1. Report of the United Nations Conference on Environment and Development (1993). UN, New York, v. III, p. 43.
2. Ibid., v. I, p. 9.
3. NATO Science and Society Newsletter (1993). № 38.

♦

SCIENCE POLICY: NEW MECHANISMS FOR SCIENTIFIC COLLABORATION BETWEEN EAST AND WEST - UNESCO'S ROLE AND POSSIBLE CONTRIBUTIONS

F.W. EDER
UNESCO, Division of Earth Sciences, Paris, France

ABSTRACT. The role and possible contribution of UNESCO in science policy is described. Proposals are given as to co-ordination and co-operation in sciences and environmental management. International studies and projects initiated by UNESCO contribute to mutually beneficial results.

Keywords: UNESCO's contribution / priority collaborative research topics / information exchange.

1. Introduction - the Role of UNESCO

The fall of the Berlin Wall has become a symbol for the rapid and complex changes which we all have breathlessly watched over the past few years.

And, as a German now serving in UNESCO, I would like to thank the Russians especially for their efforts in ending the cold war.

Today, we meet to discuss new modes, new challenges for scientific collaboration between East and West - and we celebrate freedom, I believe, because it is the only valid answer - or basis for finding answers - to the pressing issues of the post-cold war era.

During the recent general conference of UNESCO our director general, F. Mayor, emphasized that "Freedom is the content, democracy is the context and tolerance is the commitment that will give us the answers to our contemporary challenges".

That is why I am personally honoured to participate in this exceptional meeting. I am privileged to be here on behalf of UNESCO, its secretariat and member states, which have worked so effectively in returning the organization to its constitutional mission of fostering the "free flow of ideas and sciences by word and image".

UNESCO's founders, those who framed its constitution as the second world war was ending, believed fiercely and eloquently in free inquiry. They saw it as the motor of advancement in the sciences, the humanities and arts and in education. Intellectual freedom cannot exist in the absence of the free expression of ideas. And so UNESCO had to be involved in communications as a guarantor of "free flow" within and between countries. The cold war's passions and polarization threatened to compromise that commitment and it was therefore essential to return UNESCO to its ethical roots - to ensure that freedom and pluralism would guide its policies in the fields of science, education and communication.

V A Koptyug and J Klerkx (eds), Science Policy New Mechanisms for Scientific Collaboration between East and West, 19–26

In the October 1993 issue of UNESCO's publication "Sources" the question was raised "what comes after communism?". Democracy it is hoped. But the labour pains are horrible, as recent events in Eastern Europe and in other parts of the former Soviet Union have demonstrated. Economic disturbance, a decline in standards of living and political instability plague the region, dashing hopes and sowing the seeds of discontent. What role can an organization like UNESCO play in this delicate transition phase?

For a start, it can help these countries overcome the effects of a succession of disasters such as Chernobyl, or assist prestigious institutions like the Hermitage Museum, or the Bolshoi Theatre to overcome their difficulties. It can focus attention on areas until recently ignored, such as the training of environmental scientists, and contribute to the reconstruction of entire sectors such as national education systems. And, the gravity of the situation in certain countries, especially the so called "brain drain" phenomenon, is such that international aid is not only justified, but absolutely necessary.

2. Priority Collaborative Research Topics, UNESCO's Proposals

The preservation of the environment must become an absolute priority in all Eastern European countries, especially now that they have started a root and branch reconstruction of their economies. For example, in Romania, the natural assets, which are the basis for development, have deteriorated and decayed: the disfunctioning structure of the economy have led to severe pollution and overexploitation of resources. Additionally, the enlargement of areas under cultivation, the construction of dams, and deforestation have drastically disrupted ecosystems. In this period of transition, this deterioration must first be checked (monitored), then, gradually, the damaged areas can be regenerated and some improvement in people's quality of life achieved.

What are the possible specific scientific contributions of UNESCO? I would like to inform you about the outcome of our recent General Conference.

2.1. CO-ORDINATION AND CO-OPERATION FOR THE FOLLOW-UP OF THE UNITED NATIONS CONFERENCE ON ENVIRONMENT AND DEVELOPMENT (UNCED RIO 1992)

As the first point in addressing co-ordination and co-operation for the follow-up to the United Nations Conference on Environment and Development (UNCED), I would like to inform you that the four Chairpersons of the international and intergovernmental environmental science undertakings of UNESCO, the chairpersons of MAB, IHP, IGCP and IOC prepared a joint statement which was welcomed as the first step in the process of increasing co-operation and co-ordination.

The general sentiment of UNESCO's General Conference was that this coming together of the environmental science undertakings was very important, especially within the framework of the follow-up of UNCED. Many delegates endorsed the joint statement, which they considered to be an important step towards setting the stage for further action by the secretariat of these undertakings at UNESCO in Paris, as well as by national and regional committees and bodies.

Four priority areas have been identified for co-operation: coastal zones, islands, biodiversity and natural disaster reduction.

Attention will be drawn to the important role that UNESCO is expected to play in the implementation of Agenda 21, the action plan of the Rio Conference. There are high expectations of UNESCO in this respect, in particular, in the areas of science (e.g. waste disposal management), and education, but also with regard to greater interdisciplinarity in environment and development work involving all areas of expertise of the Organization.

Furthermore, the discussion of education and information questions was taken up in a joint session to examine the new Interdisciplinary and Interagency Co-operation Project on Environment and Population Education and Information for Human Development.

2.2. CO-OPERATION IN EARTH SCIENCES; RATIONAL USE OF NON-RENEWABLE RESOURCES; NATURAL DISASTER REDUCTION

A large number of speakers referred to sub-programme "Co-operation in earth sciences; rational use of non-renewable resources; natural disaster reduction". The role of earth sciences in the management of natural resources and the environment was acknowledged.

Support was voiced for the International Geological Correlation Programme (IGCP) as an important means of facilitating international co-operation in research on geological problems and of promoting a wise use of the Earth. Further strengthening of interdisciplinarity in IGCP projects was advocated. A number of delegates wished that the Programme would expand its coverage to their region.

The contribution of earth sciences to global geoenvironmental activities, and to the implementation of UNCED's Agenda-21 was recognized. The role of earth sciences in monitoring and analyzing environmental parameters, and in the exploration and use of non-renewable resources was emphasized. There was particular support for activities and systems favouring training, mapping, data handling and waste management. The prospect of selecting geological sites for inclusion in the convention on the protection of cultural and natural heritage, was well received.

The participation of UNESCO in the International Decade for Natural Disaster Reduction (IDNDR) and in the study of natural hazards of both geological and hydrometeorological origin was widely encouraged. References were made in favour of the forthcoming World Conference on Natural Disaster Reduction. The activities of existing UNESCO-related centres on earthquakes and other hazards were underlined. Delegates were also informed of plans for a new UNESCO centre related to the integrated analysis of natural disaster reduction and environment and development.

2.3. TERRESTRIAL ECOSYSTEMS: CONSERVATION AND MANAGEMENT OF THEIR RESOURCES (MAN AND THE BIOSPHERE — MAB)

Many delegates stressed the relevance of the Man and Biosphere Programme in contributing to implement UNCED's "Agenda 21" and the Convention on Biological Diversity. Several delegates drew attention to the four priority areas — biodiversity, sustainable development, capacity building and information/ communication — recommended by the MAB International Coordinating Council. Several delegates referred to MAB activities of special importance to them, such as exploring approaches to sustainable development through regional units, improving human-environment interactions in coastal areas and small islands, and better understanding land-inland water ecotones.

Concerning biodiversity, it was noted that a Forum on Biodiversity, Science and Development, organized by IUBS with the sponsorship of UNESCO and ICSU will be held at UNESCO Headquarters in 1994. Furthermore, the several delegates stressed the importance of the cooperative effort under the "DIVERSITAS" Programme of UNESCO/IUBS/SCOPE.

Biosphere reserves received much support from delegates and several mentioned that biosphere reserves are being increasingly used as instruments for national conservation and sustainable development strategies. A good number of delegates endorsed the proposal that a statutory framework should be elaborated for the international biosphere reserve network of MAB and welcomed the offer of Spain to host a second international conference of experts on biosphere reserves in 1995.

The role of MAB in developing innovative approaches to sustainable development in different bio-climatic and socio-economic situation was supported by many delegates.

The need for MAB to promote the use of modern informations systems such as GIS for ecological research and monitoring was stressed by some delegates, as was the need to further enhance capacity building, for example through creation of a training unit on forestry in Central Africa, or the new Cooperation Programme for the humid tropics.

2.4. INTERGOVERNMENTAL OCEANOGRAPHIC COMMISSION (IOC) AND MARINE SCIENCE RELATED ISSUES

Many Member States expressed strong support for the IOC programmes and firmly endorsed the decisions made by the Seventeenth Assembly of the IOC, including proposals leading to a strengthened capacity of the Commission to act as the marine science arm of UNESCO, and underlined the importance of the IOC's relationship with its partners in the UN system. Increased synchronisation between all UNESCO marine science activities was welcomed and further progress in that direction called for.

Several statements were made in support of the need to ensure and increase the status and visibility of the IOC, as well as the dependable resources required for adequate programme implementation at the regional and global levels.

The important and leading role of the IOC in follow-up to UNCED was commended and strongly supported. Particular reference was made to the IOC role in global climate change studies, inter-agency co-operation, development of the Global Ocean Observing System (GOOS) and related capacity building in developing countries.

Several delegates endorsed the work of the Commission with respect to tsunamis and requested that it consider activities related to the study of storm surges, tropical cyclones and hurricanes, especially from the perspective of developing early warning system for these natural hazards. Other delegates requested help in strengthening marine and ocean research institutes in their countries.

Most speakers expressed support for declaring 1998 the International Year of the Ocean as proposed in document 27 C/99 (Agenda item 6.9).

2.5. HYDROLOGY AND WATER RESOURCES MANAGEMENT (INTERNATIONAL HYDROLOGICAL PROGRAMME — IHP)

In the field of hydrology and water resources management, a great deal of support was given to IHP which was recognized as one of the major programmes of the Science Sector; closed coordination with WMO's Operational Hydrology Programme (OHP) was recommended in accordance with the recommendations of the UNESCO/WMO/ICSU International Conference on Hydrology.

Several countries recommended the strengthening of the Regional Offices to enable greater development of hydrological activities at the regional level.

Special interest was expressed in several topics within the IHP such as freshwater resources, climate change, small islands, humid tropics, water quality, pollution problems and karst hydrology. Furthermore, several countries expressed concern over the problems of arid zones and recurring droughts and called for increased research on the matter, particularly in Africa. Attention was also drawn to the dramatic situation of the Aral Sea and its negative impact on the environment and UNESCO was requested to the alleviation of this situation. Considerable interest was expressed in the Blue Danube project and in the development of research on the impact of the Chernobyl accident on water resources.

A great number of delegates emphasized the need to give education and training priority within IHP, including the aspect related to informing the general public, and strong support was expressed for the continuation of IHP with particular reference to capacity building.

2.6. CO-OPERATION IN BASIC AND ENGINEERING SCIENCES

The majority of delegates gave high priority to programme II.1 "Science and Technology for Development", and its subprogramme "Co-operation in Basic and Engineering Sciences". They expressed their satisfaction over the actions proposed and efforts for concentration on these actions made in the programme. Many delegates stressed the need to focus on the development of the countries human resources in the basic and engineering sciences. Basic and applied research capacities need to be built up through the organization of advanced training in mathematics, physics, chemistry, biology sciences and engineering sciences.

The need for increased efforts in the improvement of university teaching in the basic sciences was recognized by many delegates, and co-operation between universities within the framework of UNITWIN and the UNESCO Chairs programme was highly appreciated.

Pilot projects aimed at the training of university science teachers, at curricula development, and at the preparation of a new generation of distance-learning programmes, received broad support. The creation of fellowship programmes to be organized in developing countries should be encouraged within the framework of technical co-operation among developing countries.

Ten delegates called for strengthening the programme of mathematical sciences and, in this respect, it was pointed out that optimal use should be made of International Centre for Pure and Applied Mathematics, for which the support of the French Government was acknowledged.

Many delegates endorsed the proposal to transfer the International Centre for Theoretical Physics (ICTP) to UNESCO, and highlighted ICTP's unique capacity to provide high quality advanced training in physics to thousands of specialists, particularly those from developing countries. Several delegates stressed the need to ensure the functional and intellectual autonomy of ICTP. Some of them insisted on the fact that the transfer of ICTP should not have additional financial implications to UNESCO in the 27 C/5. The longstanding Italian contribution to ICTP was gratefully acknowledged.

With reference to the field of chemistry, several delegates requested UNESCO to support the 6th International Conference on Chemistry, the African Union of Pure and Applied Chemistry, the International Centre for Phytochemistry and the International Programme for Chemistry for Life.

Several delegates called for the further development of major UNESCO projects in modern biology. Positive references were made to the UNESCO/ICRO training and fellowship programmes in cellular biology, the Molecular and Cellular Biology Network, UNESCO/IBRO co-operation in neurobiology and the Botany 2000 programme. Many delegates underlined the priority of activities in biotechnology and microbiology and supported the MIRCEN programme.

Several delegates emphasized the importance of activities related to the human genome and molecular genetics. Mention was also made of the need to increase the involvement of scientists in the discussion and policy-setting on bioethics, and especially on the ethical implications of human genome research.

The majority of delegates emphasized their satisfaction and support of the ecotechnologies theme, particularly solar energy. They expressed their appreciation of the organization of the high-level expert meeting of the World Solar Summit Process, held at UNESCO Headquarters this year. Several delegates urged UNESCO to support Member States in organizing various regional meetings on the subject of solar energy. This process might lead to the organization of Summit of Heads of States.

Twenty-three delegates made special reference to the launching of UNISPAR, a new programme created to facilitate and intensify co-operation between the university, science and industry, a process seen as essential to achieving sustainable socioeconomic development.

Several countries in Africa supported the establishment of the International Fund for Technology Development in Africa. A number of delegates from Latin America mentioned the Bolivar programme and urged UNESCO to continue to actively participate in this programme.

Delegates also stressed the importance of strengthening the following international and regional networks in engineering sciences: the International Network of Earthquake Engineering Centres, the World Solar Summit Network, the Common Market of Knowledge, the Engineering Deans and Industry Leaders Network, the Continuing Engineering Education Network, and the African Network of Scientific and Technology Institutions (ANSTI).

A significant number of delegates expressed concern regarding the future role the theme "Science, Technology and Society" will have in UNESCO's programmes, especially the importance of facilitating the incorporation of science and technology into national socioeconomic development planning, and promoting science throughout modern society.

In this respect, the representative of the Director-General indicated that the Organization will continue to co-operate with Member States on these issues, within the framework of the Participation Programme, and that an addition point will be interested into the principal resolution on Major Programme Area II, referring to training in the management of science and technology and in information systems.

Many delegates from developing countries underlined the importance of UNESCO's assistance in the field of science and technology policy and management, and stressed the need for efficient relationships between the scientific community and socioeconomic decision markers.

Some delegates from the African region stressed the need to follow up the recommendations of CASTAFRICA II. Many from developing countries stressed the important role regional networks such as STEPAN play in strengthening indigenous capacity in the field of science and technology policy and management.

3. Exchange of Scientific Information

In this context the Science Sector of UNESCO has developed activities to support democratic reforms in the countries of Central and Eastern Europe along the following lines: creating an appropriate scientific and technological base for a market-based economy; helping the transition from an armament-oriented technology to a modern civil industrial infrastructure; improving scientific information exchange and international networking among scientists (both individuals and professional organizations); and encouraging new approaches and new problems of thinking, as well as development of concrete activities related, *inter alia*, to the protection of the environment.

The project on "Transformation of management system of science and technology in the countries of Central and Eastern Europe during the transition to market economy" has been carried out in 1992—1993 under the "Science, Technology and Society" programme. Within the framework of this project, a study with policy recommendation was prepared by a group of Western and Eastern European experts. The study was discussed at a meeting in Varna (Bulgaria) in June 1992. One of the recommendations of the study calls for supporting and fostering collective public awareness of the results of scientific and technological activities and the use made of them by society, while keeping this type of activity under enlightened democratic supervision.

The floating university project executed jointly by UNESCO and the European Science Foundation (ESF), enjoined multilateral co-operation with countries of Europe, North Africa and Latin America. Three research vessels of the Russian Federation have been working in the Mediterranean, Black and Baltic Seas in 1993. Thirty five students from the Russian Federation and Ukraine have been trained.

The role of governments and parliaments in the reorganization of science in Europe and the establishment of new scientific and technological legislation were discussed at international meetings organized by UNESCO in Venice (Italy) in April 1992, Manchester (United Kingdom) in April 1993, and Siena (Italy) in May 1993. The first phase of the project on the Safeguarding of Russian Basic Science was implemented in co-operation with the Russian Academy of Sciences in 1992. Emergency assistance was provided to the Bulgarian Academy of Sciences and agreement on co-operation was prepared and signed with the Rumanian Academy of Sciences.

A Conference on the Role of Science in Rebuilding Central and Eastern Europe, co-organized by the International Council of Scientific Unions (ICSU), UNESCO and the Commission of the European Communities in Leeds Castle, Kent (United Kingdom) in April 1993, gathered participants from 15 countries, including a number of leading scientists, ministers, parliamentarians and policy-makers. The purpose of the conference was to draw attention to the serious problems facing science in Central and Eastern Europe and to explore strategies for solutions. Participants were unanimous in stressing the important role of science and technology both in achieving economic progress and prosperity, and as part of a nation's cultural heritage. It was also stressed that adequate measures should be taken by governments to safeguard science since it is very difficult to re-establish the necessary human resources and infrastructure once these have been dissipated. The conference adopted a number of recommendations, including increased provision of training and educational opportunities for young scientists in their own countries and abroad, increased research on technology transfer between East and West and between basic research and industry.

A series of national studies was initiated, and the international meetings devoted to the growing problem of scientific brain drain from the countries of Eastern and Central Europe and the former Soviet Union were organized in Moscow (Russian Federation) in February 1992, Venice (Italy) in October 1992 and Warsaw (Poland) in November 1992. The strategy to combat the brain-drain phenomenon and to organize a mutually beneficial exchange of scientists between East and West European countries is under discussion. UNESCO has also contributed to the fund for emergency assistance to the Russian Academy of Sciences.

4. Modes of Realization – Proceed

In late 1992 the Programme for Central and Eastern European Development (PROCEED) was established.

PROCEED is designed to co-ordinate UNESCO's activities in support of Central and Eastern Europe and in the newly independent States of the former Soviet Union, both within and across the Sectors. To do so, it focuses on the priority needs of the region and its sub-regions, in particular: the re-establishment and consolidation of democracy; pluralism in beliefs, attitudes, institutions; the expression of identity, and protection of minorities and human rights; preservation of the environment and of the cultural heritage; human resource development, especially in the areas of UNESCO's mandate reinforcing long-term independence and durability.

The implementation of PROCEED activities is the responsibility of the programme sectors but co-ordination is entrusted to the PROCEED Unit which also acts as a focal point for Member States and external agencies. PROCEED co-operates closely with the UNESCO Office recently set up in Moscow to co-ordinate activities within the Russian Federation, with the Vienna and Geneva Liaison Offices (both of which maintain a regular dialogue with other United Nations agencies), and with the sectoral offices in Bucharest (European Centre for Higher Education — CEPES) and in Venice (Regional Office of Science and Technology for Europe — ROSTE).

PROCEED is financial from both UNESCO regular programme and from extra-budgetary resources (such as fund-in-trust arrangements with bilateral government donors and other sources, UNDP, European regional funding mechanisms, foundations and the private sector).

♦

RESEARCH PROGRAMS OF THE COUNCIL OF EUROPE AND POSSIBLE WAYS OF COLLABORATION WITH SCIENTISTS FROM EASTERN COUNTRIES

J.P. MASSUE
European Federation of Scientific Networks, Council of Europe, Strasbourg, France

ABSTRACT. Possible ways of collaboration between European and the FSU countries are discussed. The important role of networking is underlined and programs of cooperation presented.

Keywords: collaboration / scientific networks / training.

This paper intends to discuss three important problems. The first one is dealing with the idea of the importance of networking for scientific cooperation. Secondly, I would like to give you some information about the existing program to support scientific research, in the framework of the European programs, and, thirdly, I shall present you a specific European agreement in which Russia is a full member state.

A network is a means, a tool, and setting up of networks facilitates joint projects of research for training or technological development. The main idea is to give to the scientific community full responsibility in setting up scientific networks in different areas, the network being an initiative of scientists. It means that this process should be a "bottom-up process", i.e., not a network set up by the administration of different ministries in charge of research, but more an initiative coming from the scientific community.

This idea facilitates cooperative actions for research and technological development. On the basis of this initiative, a number of networks have been set up in many different areas: in the field of culture, we have a network of science and technology applied for the protection of cultural heritage; we also have a network dealing with material sciences. On the basis of those different networks set up by the initiative of the scientific community, it was decided to create the European Federation of the European Scientific Networks. The European Federation was set up this year, on the 26th of January; it is under the umbrella, that has strong links with UNESCO, with the Commission of the European Community, and the Council of Europe. We shall discuss one concrete example of a network in the field of material sciences later on. Now, I would like to offer some brief information about different programs at the Community level to support scientific research open for the former Soviet Union.

The first program is a technical assistance program for the New Independent States, what we call the "TACIS" program. On the basis of the ground assessment of the economic situation and the reform process in the former USSR, the European Community and the

V A Koptyug and J Klerkx (eds), Science Policy New Mechanisms for Scientific Collaboration between East and West, 27–30

authorities of the former USSR decided to concentrate in 1991 the technical assistance program in the following five sectors: private sector training, energy, transport, financial services, and food distribution. In October of 1991, the member States approved the first proposal for the energy sector of 115 million Ecus, then for other four sectors programs, and training - 100 million Ecus. In 1992, the TACIS program was launched and covered the following broader areas: support for enterprises, human resources development, food production and distribution, network. In addition to the national program, the 1992 TACIS program amounted to 29 million Ecus to fund regional interstate program.

Following this, and in line with the new regulations provision, the Commission prepared constitutional coordination with the authorities of the New Independent States, an indicative program for technical cooperation with these countries over the period of 1993-95. The Independent States and the indicative program received a favorable opinion of the TACIS management committee in June 1993. For a good idea of the funds available, one can mention that in 1992 the total sum was amounted to 450 million Ecus, for 1993 up to 510 million Ecus. It was decided for 1993, and will be followed in 1994, to add a selection of geographical concentration areas. From the 1993 budget action programs, two new areas of geographical concentration are prepared: the Urals with Chelyabinsk, Perm, Ekaterinburg, and West Siberia with Kemerovo region, and Novosibirsk. Novosibirsk is a major center for scientific expertise, again largely within the military industrial complex. These areas, the Urals and West Siberia, represent the core of the problem of transition, particularly in the Russian Federation. The TACIS program also has many different sectoral programs. I would like to underline that for the TACIS program the role of the coordination units is very important. And proposals for the program have to go through the coordination unit.

What are the other programs open for cooperation with Russia? The present situation, in 1993, is the following. The funded cooperation in 1993 is through INTAS - the International Association for the Promotion of Cooperation with Scientists from the Independent States of the former Soviet Union. Fifty five projects have already been approved within the scope of this program, forming the core for the declaration of intent launched in mid-September 1993 and closing on 16th October 1993. 3800 requests for funding collaborative R&D projects in all exact sciences were received. The available budget at the moment is only 21 million Ecus. Anyway, this is something, and the R&D program is launched. Five of the 15 specific programs of the above mentioned follow-up program for research and technological development are open for participation. "Open for participation" means that one can cooperate with labs from the Community countries in a specific program set up by the Commission, but without financial support. The projects open for participation are in the fields of the environment, non-nuclear energy, nuclear fission safety, biomedical and health research, but there is no funding available at the moment for new independent state based organizations.

Eurocap participation is possible, too. The Russian Federation is already involved in several, and its application for membership was in principle accepted in June 1993. With the nuclear fission safety program, the European Community has set up a laboratory in Chernobyl. Extension of the collaboration with Russia and Belarus is under discussion. In the field of energy, DG-18 has set up a number of energy centers in the contacts of training programs which end with TACIS funding. One can enter in this program through TACIS. ISTC - International Science and Technology Center - within TACIS for the conversion of military scientists is not yet operational. All I have mentioned is still valid for 1994. But it is planned that, other than INTAS, funding for cooperation with the New Independent States based organization will be possible through two mechanisms provided the requested HABAS funding is granted. First, participation in five open specific programs of the framework program. Secondly, some limited funding for the new independent state organization to participate in R&D collaborative project between the European Community and Central and East European countries. A special call for proposals is expected to be announced in December.

The fourth framework program (that is, between 1994 and 1998) will take effect. All R&D cooperation with the New Independent States will be funded and handled through it. Further specific collaboration projects will probably be developed in these contracts.

How is it possible to start cooperation? Let us take an example from the area of material science. Material science is a very important area for research and technological development. It covers interdisciplinary areas, and you have the application in all fields of industry, transport, microelectronics, biomaterials and so forth. And in the Western part of Europe, we have set up the European Material Research Society (EMRS). Twice a year we organize meetings in Strasbourg, and we have developed different sectoral sub-networks. The interest of this EMRS network is that when the opportunity for research and training arise, e.g. in the framework of the Community programs, we can react immediately, having a network of labs ready to respond. This is something very important for the research development of training programs. What's more, thanks to UNESCO we have set up at the international level an international committee of material research society. And we have links with the United States Material Research Society, with Japan, with China, and I think that it will be the right moment for Russia to set up the Russian Material Research Society. This Society could be built on four subjects and will operate one for the Novosibirsk area, one for the Urals, one for the Moscow area, and one for the St. Petersburg area.

About the European program where Russia is a full member state. In 1987 we set up a European open partial agreement in the field of major disasters (major hazards - natural and technological). The idea was to open the door to the Eastern European countries. That is why we have called this agreement an open and partial European agreement. It is partial because not all West European countries are taking part in this agreement. Only the interested ones are taking part. But it is an open agreement, because this agreement was supposed to open the door to the East European countries. In 1990 the former USSR became a full member state of this European agreement. Now the situation has changed, and we have Russia, Ukraine, Belarus, Armenia, Georgia, Azerbaijan, and other states in this European agreement. It is interesting, because thanks to this program we have the possibility also to facilitate the training and research program for the prevention and protection against major hazards.

Another concrete example is linked with training, because here, we have a possibility to have very close cooperation. We decided to launch a new European program for training in the field of risk sciences. The field of risk sciences is something new, a new interdisciplinary concept, and our area is the following. We benefit first of all from the 15 year experience of a German university - the University of Wupperthal where they have developed a curriculum for training students in the field of risk sciences, but more oriented toward the problem of risk in underground work conditions, and this is a large part of the field of risk sciences. They have a training program from the first year of university and to the Ph.D. level. At the beginning only 50 students studied at this university, and by now they have had 1500 students in five years.

Therefore we decided to start a project in the field of risk sciences.

The purpose of this project is to have a kind of cartography of various universities and research centers. Having the interested institutions, we shall adopt the same pattern for the training and propose specific training models for training. It implies that a student interested in the area of risk sciences will have the cartography with information on all universities and training institutions in general, with an information to help him select training models he can follow in different universities. The end of our project will be to build a mobility program for students and lecturers based on the same educational concept. It must be a very concrete, specific program. And as the field of risk sciences is interdisciplinary, it is very interesting for the people interested in the training and cooperation for training at the European level.

To summarize, my first proposal is to consider this idea and take the initiative in the field of material science, and the second to take part in this European program dealing with training in the field of risk sciences.

EAST-WEST SCIENTIFIC COLLABORATION - SOME IDEAS FROM THE LABORATORY

M. ANDRE
Centre d'Etudes Nucléaires de Cadarache, Saint-Paul-Lez-Durance, France

ABSTRACT. East-West scientific collaboration today needs to be encouraged by actions able to increase mutual knowledge. Problems of collaboration, including financial support, are considered and suggestions made.

Keywords: east-west collaboration / reciprocal knowledge / support.

In the western world, and particularly in political circles, everybody agrees about the necessity to consolidate collaboration with Eastern countries and to avoid dismantling scientific institutions. At the governmental level, or through the EC, credits are planned and given but their benefits are long to be seen. Regarding the scientific field, contacts that we can make during international conferences show that our eastern colleagues are a bit disappointed.

Of course exceptions can be noted, for instance in space and nuclear research. Hence, collaborations with Russia are taking place at Cadarach for the safety of nuclear power plants, the breeders and on radio-ecology. In the most current domains, only national institutions, governments and international authorities can promote East-West cooperation by significant budgetary incentives. We must notice that local research units usually face budgetary problems, whatever the absolute value of those budgets. At that level, because of international competition in science, motivations can practically come from short term interests in having a partner to exchange services, or to plan complementary approaches to research. A serious incitement can follow from the obligation to have eastern partners in order to obtain research contacts, as it is usual for the contracts inside the EC. But if such an active policy is internationally undertaken, everything will depend on the knowledge and on the competence of the partners.

Therefore, after discussions with some colleagues, suggestions could be discussed in order to facilitate "natural" relationships between East and West and to encourage collaborations. Actions, perhaps symbolical compared with the dimension of the problems, can be sought after, able to increase the mutual knowledge, very imperfect after all those decades of total isolation. As it seems, it is a primary condition to set up real co-operation which always assumes reciprocal knowledge, esteem and trust.

We will develop the latter conditions before evoking any problems of means.

V. A. Koptyug and J. Klerkx (eds.), Science Policy: New Mechanisms for Scientific Collaboration between East and West, 31–34.

1. Information

1.1. RECIPROCAL KNOWLEDGE IS OBTAINED THROUGH THE ACTIVE DIFFUSION OF INFORMATION

As far as I am concerned, I have discovered the existence of the numerous Siberian Research Centres, (with the exception of the Centre for Closed Ecological Systems at Krasnoyarsk), only by the present invitation. In France, Research is mainly organized in National Institutes as INRA (Institut National Agronomique) or CNRS (Centre National de le Recherche Scientifique) or INSERM (Institut National de la Sante et de la Recherch Medicale) and the "Annuaires" are published each year comprising lists of laboratories with related themes of studies. They can be obtained on request. Similar information should exist for the other western and eastern countries: it could be useful for everybody to clarify that point and to discuss how to obtain those documents. Naturally, the help of scientific services of embassies may be requested; it seems that way is not sufficiently used by scientists.

1.2. SCIENTIFIC SOCIETIES

Membership in Scientific Societies systematically allows the availability of information on the life of a particular branch of research: events, national conference or workshops... Later on means will be proposed to reinforce this participation.

1.3. SCIENTIFIC NETWORKS

It is relatively simple and inexpensive to create scientific networks in order to make circulate the information. These networks concern a limited field of research, for instance the adaption to drought. They can be efficient, rather cheap, being only exchanges of information in the form of letter or bulletin. They should encourage the collaboration and avoid the duplication of research.

2. Scientific publications

2.1. LANGUAGE PROBLEMS

It must be acknowledged that establishing a long term collaboration is based on quality and efficiency of research, and on the complementary evaluation reciprocally acknowledged by the partners. The nature, the quality and the number of papers issued from the laboratories can give this evaluation. Here we come up against a problem of lack of knowledge due to the political isolation and the closure of borders in the past. It should disappear now. The problem of national language remains in the international context. It is generally agreed that publishing in international journals, in English, is recommended. But it is not the only solution because it leads to depreciate national languages that are the basis of cultures. The richness of our communities is built on our diversity. With the standardisation of languages and cultures, comes the risk of impoverishment. Better than publishing systematically in English, the best national journals could be recommended to make easier the access to papers in national language for foreigners with some adjustments. It would be judicious to add translations of summary, legends of figures and tables in English in national journals. In those conditions, the originality and the interest of all papers could be evaluated and possibly translated and cited.

2.2. EXCHANGE OF PUBLICATIONS AND ANNUAL REPORTS

In spite of the difficulty that the language barrier represents, the exchange of papers from each laboratory wishing to cooperate is always desired. To follow that advice, I will add some typical references on ecophysiology from our Department [1—7]. The diffusion of annual reports is also recommended in that context; they automatically include the references of publications.

3. Developing the mutual knowledge: Twinning?

3.1. VISITS, EXCHANGES (AS A MATTER OF INTEREST)

3.2. TWINNINGS

A good collaboration can be established and can last only if, besides the knowledge and recognition of capabilities, there are systematic visits and personal relationships establishing a climate of trust. For that aim, twinning between centres, Institutes or even departments seems fit. It is through the twinning of cities that a better understanding between European citizens has been achieved since World War Two. It is an incentive to come out of isolation, to surpass cultural and political borders. Is it an idea to be extended to the scientific field?

4. Means of collaboration

4.1. SUPPORT OF THE ACTIVITIES OF MUTUAL KNOWLEDGE

A part of the means depends on the personal involvement of the scientists and of people in charge. Visits and twinning, for example, can be developed, like many networks, from voluntary activities having no particular budgets. However, this development will remain uncertain, slow and probably limited if there is no official incentive through financial support (long distance travelling, problem of currencies). As said above, helps to mutual knowledge and direct relationships should anticipate and accompany a policy of international collaboration.

Scientific Societies. A maximum of reciprocal memberships should be recommended with special tariffs for eastern scientists because the exchange rate is very bad for them nowadays. Several scientific unions, like the American Physical Society, take care of that problem for their membership and the subscription to their journals. In order respects, the life of the societies and of the associated scientific journals is often precarious compared with the American ones, concerning a wider scientific population. Can we imagine that some grants, coming from international authorities (NATO, EEC), could be given to Western journals proportionally to the number of their Eastern subscribers?

Support to Visits, Twinning (as a matter of interest).

4.2. DIRECT SUPPORT TO COLLABORATION

4.2.1 *Proposals for National and International Programs* (as a matter of interest).

4.2.2 *Exchange of Scientists and Students* (as a matter of interest).

4.2.3 *Contracts to Purchase Equipment or Services.* Naturally, commercial activities between Eastern and Western companies are not new. But let us imagine that a small Western laboratory wants to buy a device from an Eastern university. After requesting a proposal price, are we certain to go through administrative procedures simply and quickly enough, in comparison with current practice, in order not to discourage this purchase?

4.2.4 *Exchanges of Services and Equipment with no Money.* The present exchange rate of certain currencies can make the conclusion of contract between laboratories very problematic. In the context of already existing contacts between laboratories with personal relationships implying a certain confidence (twinning), can we imagine collaboration set up on exchanges with no money involved? For instance exchange of services such as software or models for equipment or hardware? In what measure, with all required prudence, can we evoke that possibility of exchange, in so far as a compensation in services, studies, transfer of know-how can be considered from our colleagues? What procedures can be anticipated at the administrative level?

References

1. André M., Daguenet A., Massimino D., Gerbaud A. (1985). The $C_2$3A System, an Example of Quantitative Control of Plant Growth Associated with a Data Base. In "Proceeding of Fifteenth Intersociety Conference on Environmental System" San Francisco Calif. July 15—17, 1985. Publisher: SAE The Engineering Resource for Advancing Mobility, Paper №851395 et Rapport NASA TM 88215.
2. Morin F., André M., Betsche Th. (1992). Growth Kinetics, Carbohydrate and Leaf Phosphate Content of Clover (Trifolium Subterraneum L.) after Transfer to a High CO_2-Atmosphere or to High Light and Ambient Air, Plant Physiol. 99, 89—95.
3. Kent S., André M., Cournac L., Farineau J. (1992). An Integrated Model for the Determination of the Rubisco Specificity Factor, Respiration in the Light and other Photosynthetc Parameters of C_3 Plants *in situ*, Plant Physiol. Biochem., 30(6), 625—637.
4. André M., Cloux H. Du. (1993). Interaction of CO_2 Enrichment and Water Limitations on Photosynthesis and Water-Efficiency in Wheat. Plant Physiol. Biochem. 31(1), 103—112.
5. André M., Thiery J., Cournac L. (1992). Ecosimp2 Model — Prediction of CO_2 Concentration Changes and Carbon Status in Closed Ecosystems. Communication at COSPAR-World Space Congress, 18 August—5 September 1992, Washington D.C. in press in Adv. Space Res.
6. Chavardieff P., d'Aletto T., André M. (1992). Specific Effects of CO_2 Concentration Doubling on Productivity and Mineral Content in Lettuce. Communication at COSPAR-World Space Congress, 18 August—5 September 1992, Washington D.C. in press in Adv. Space Res.
7. André M., Peltier G., Chagrvardieff P., (1993). Ecophysiology in Closed Systems — From Closed Growth Chamber to Artificial Ecosystems. Proceedings of the Int. Committee of Closed Ecology Experimental Facilities, 22—25 March 1993, Aomori, Japan.

♦

II. EXPERIENCE OF THE INTERNATIONAL RESEARCH CENTRES' ACTIVITIES AND SOME RESULTS OF INTERACTION BETWEEN SCIENTISTS IN PRIORITY AND MULTIDISCIPLINARY PROBLEMS

II.1. ENVIRONMENTAL PROBLEMS

INTERNATIONAL COOPERATION IN BOREAL FORESTS STUDY OF SIBERIA

E.A. VAGANOV
Siberian International Centre for Ecological Research of Boreal Forests, Sukhachev Institute of Forest of SB RAS, Krasnoyarsk, Russia

A.S. ISAEV
International Forest Institute, Moscow, Russia

S. NILSSON
International Institute of Applied Systems Analysis, Laxenburg, Austria

ABSTRACT. Fundamental researches of Siberian forests (conditions, supply, dynamics, successes) are encouraged by leading research institutions of SB RAS by the Forest Institute and Siberian International Centre for Ecological Research of Boreal Forests. These institutions coordinate a number of important international projects directed towards the analysis of forest ecosystems during global and regional changes of climate, of the role of Siberian forests in the carbon balance of the planet. Scientific institutions of forest profile from the USA, Switzerland, Canada, Finland, China take part in realization of joint projects.

Keywords: boreal forests / international cooperation / forest resources / global changes.

Siberian Forests occupy a considerable part of the Asiatic Continent. They constitute nearly 20% of the world forest resources, among them 38% of the most valuable coniferous forests. The role of Siberian forests as a biomass producer and carbon accumulator is acquiring a global significance. There are concentrated about 30 billion tons of bound carbon which in case of release may change the state of the atmosphere drastically. It is particularly impossible to assess the character of global changes in the environment and the extent of impact of these processes on the planet biosphere without detailed information on the resources and the ecological role of Siberian forests.

Preservation of ecological and resource potential of Siberian forests growing in severe climatic conditions and which are subjected to constant anthropogenic pressure is a very complex problem. It can't be solved successfully because of the deterioration of the Russian economy and an absence of development of infrastructure in most multiforested regions of Siberia and the Far East.

Changes that are taking place in Russia are opening up unprecedented opportunities and at the same time constitute serious danger to the preservation of nature and forests of Siberia and the Far East. Under such conditions it is indispensable to revise policies of the forest sector in

V. A. Koptyug and J. Klerkx (eds.), Science Policy: New Mechanisms for Scientific Collaboration between East and West, 37–44.

ecology, economy and social aspects and to choose the rational strategy of utilization of forest resources. As a result of such revision, there must be a well-constructed program of forest resources management of Siberia and investment of the forest sector which is established on support of reforms, the development of needed infrastructure and the implementation of new technologies.

In this century forest study of Siberia has started in the thirties with the commencement of industrial exploration of this extremely rich region. However, the development of planned fundamental exploration started late in the sixties in connection with the foundation of the Siberian Branch of the Academy of Sciences of the USSR and creation of a network of research institutions in different regions of Siberia.

At present Siberia possesses a huge scientific potential in the field of forest science and has a lot of experience in the study of Siberian forests. The Siberian International Centre for Ecological Research of Boreal Forests, located in the city of Krasnoyarsk was set up on the basis Institute of Forest of SB RAS — the leading Russian research institute of forestry. In over 35 years of intensive work the Institute has accomplished a number of major scientific programs on forest study of Siberia: their structure, conditions, features of development, resource and ecological potential. The results of this research are summarized in more than 100 monographs and thousands of articles dealing with the problems of forestry, ecology, biochemistry of woodplants, of forest soil science, inventory and forest management, raw-wood management, forest microbiology, forest fire protection, protection from pests and diseases and other problems.

The Centre works cooperatively with many Russian institutions dealing with forest problems: Centre of Ecology and Forest Productivity Problems of Russian Academy of Sciences (CEPF) — Moscow; The Ural Sector of the Institute of Forests of the Ural Branch RAS — Ekaterinburg; Archangelsk Institute of Forests (AFI); The Far East Institute of Forestry (DalNIILH) — Khabarovsk *et al.*

The goal of the International Centre is to form a data base for the assessment of the forest covering dynamic changes in different regions of Siberia, effect of disturbance (natural and anthropogenic); organization and scientific promotion of forest monitoring and forest protection; testing new technologies of utilization and reforestation on the basis of long-term economic projects, taking into account the natural features of forests, the establishment of international laboratories, expedition and cooperative research, based on the network of Siberian expedition stations.

The International Centre works cooperatively with foreign scientific teams and organizations. In the framework of cooperation with the Swiss Federal Institute of Forest, Snow and Landscape Research (FSLR) and Arizona State University a project is being elaborated of dendroclimatic monitoring of the boreal forests of Siberia, Canada and the United States. The project promotes the development of methods and programs of statistic and imitation modelling of forming an annual ring depending on conditions of growth and climatic constituents. The utilization of these methods makes it possible to assess the dynamics of wood biomass in different geographical regions of the boreal zone to identify the change of increment depending on major climatic parameters over the past 500 years and describe the history of Siberia in Holocene.

A number of interesting study projects are done cooperatively with the Federal Forest Service of the USA (USDA FOREST service) and its exploration stations. We can refer to the study of carbon budget in boreal forests, impact at global changes on biochemical cycles and productivity of Siberian and North American forests, elaboration of forecasting models for forest vegetation inventory, analysis of forest pests density dynamics. In cooperation with the University of Turku (Finland) biochemical aspects of interaction between trees and insects are explored. The joint Forestry programs are performed with the Institute of Geography of the Academy of Sciences of China (IGCAS), National University and Department of Forestry and Forest Production of Republic Korea.

In 1993 on the premises of the Siberian International Centre an international conference was held and a major field experiment was carried out which served as a foundation for working out the "Forest Fires in ecosystems of North EuroAsia"project.

The most large-scale project practically being completed is being performed by the International Research Team jointly with the International Institute of Applied Systems Analysis (IIASA), Austria. The project was formed on the basis of a special agreement between IIASA and Russian Academy of Sciences. The project is entitled: "Forest Resources of Siberia, their management and corresponding socio-economic and nature preservation problems". The major goals of the projects are: analysis of the state of Siberian forests, assessment of their biospheric role and effect on global climatic changes, evaluation of resource potential of the forests and generation of policies of their management, which promote a sustainable socio-economic development of Eastern Regions of Russia.

The work on this major project has started but the scope is rigidly limited by financial restrictions. At present a data base is being formed on forest resources of Siberia, infrastructure, industrial and socio-economic factors. It is planned in 1994 to carry out work on the evaluation of ecological and resource potential of this forests and study of international and domestic wood market. A prerequisite for developing this project is a substantial financial support both inward and outward.

We think that the project is in accord with the global program for ecological aid of the World Bank "Global Environmental Facilities Program" and "Environmental Management Technical Assistant Project". It is desirable that in the realization of this project scientific structures of NATO should participate within the framework of the program "New Mechanisms for Scientific Collaboration between East and West".

The analytical structure of the project is indicated in Table 1. In 1994 the work of blocks 0—1 will be completed (generation of data base and pre-feasibility of existing data). Assessment studies completing the analysis and build up of scientific technical policies are planned to be carried out in 1995—1996.

The accomplishment of the project required the organization of new forms of international collaboration and the establishment of a wide network of Russian research structures involved in the project.More than 80 specialists from research and project institutes of Russia are participating in the generating of data base and information analysis but they are engaged not only in collecting information but are taking part in an active creative analytical experiment. A significant role in the organization of the project is played by IIASA as a specific methodological and organizational centre of international cooperation. In the framework of IIASA scientists from Finland, Sweden, Germany, Italy, USA, Canada and South Korea are working in the Siberian project. The fulfillment of the project has two aims: the extension of international cooperation on the basis of utilization of study results of Russian scientists in the global assessment of resource and biospheric functions of Siberian forest and the creation of sustainable conditions for incorporating international experience in the solution of concrete problems of socio-economic development of Siberian regions under the conditions of transition to a market economy. These aims also determine the formation of two interrelated trends: resource-ecological and socio-economic which in aggregate are supposed to give a scientific ground of a sustainable development of the forest complex of Siberia.

In the framework of the first trend the following subprojects are being carried out.

1. Assessment of the Resources Potential of Siberian Forests

The goal of the subproject is to form a reasonably ample and reliable data base concerning the forests of Siberia in order to provide a quantitative estimation of the resource potential of Siberia, to generate forest fund dynamics forecast and to build up reasonable strategies of inexhaustible utilization and reproduction of forest resources. The data base is set up within the existing ecological and economic regions of Siberia and includes information on the state forest inventory, materials of forest management,experimental data on phytomass and forest productivity, as well as availability of forest resources within each concrete territory.

The estimation of stock and trade quality of the wood includes definition of volumes of raw wood resources of the main forest species within each forest management enterprise "Leshoz" and integrated within each administrative territory (republic, territory, region and ecoregions). A prerequisite of this estimation is working out a new type of commodity table taking into account international (European) standards of timber in the round. It is very significant while compiling these tables to estimate the spread of defective wood in Siberian forests and make an inventory of trade quality of wood as a result of intermediate cutting.

The assessment of non-wood resources in forest ecosystems of Siberia includes the estimation of reserves of technical, food and medical raw material. It is based on an inventory of green biomass, food and medical resources of the forest, reference material and the interconnection between types of raw material and taxation characteristics of afforested areas.

While developing the taiga territories it is important to have a concept about ecological and economic accessibility of forest resources. The purpose of this analysis is to exclude from the exploitable forest fund those forest fund areas where felling is not allowed because of ecological considerations (particularly protected forests) low accessibility or the low value of stands.

The estimation of economic acceptability of forest resources is based on the rental estimation of exploitable forests and it envisages calculations of expenses and results in three aspects: harvesting, forest regeneration and transportation (wood delivery to customer). The main characteristics of the forest resources are the following: exploitable (commercial) growing stock (m^3/ha), species composition, site index, average volume of stems relief and soil conditions, distance between the forest area and customers and type of transportation. These characteristics of the forest fund determined the technology of harvesting and types of machines used in overall wood biomass harvesting and delivery to store houses.

The estimation of economic and ecological accessibility of the forest fund is regarded as an important factor of forest utilization under market conditions of sale of forest resources or lease holding or concessions.

Components Activities	Forestry	Ecology and Global Change	Industry and Infrastructure	Markets	Socio-Economics
Block 0 Pre-feasibility of Existing Data + Set-up of Study	Integrated pre-feasibility study of existing data on all components and set up of the study				
Block I Collection and Analyses of Existing Data	*Forest Resources *Forest Management *Management Technology *Economic conditions	*Description of ecological status *Global Change status	*Description of existing industrial structure (capacities and technological status) *Description of existing infrastructure	*Description of international markets *Description of domestic markets	*Description of socio-economic conditions *Description of socio-economic indicators
Block II Assessment Studies	Forest Resources Assessment *Allowable cut *Sustainable forest structure *Multiple-use *Protection *Forest management	Ecological Assessment *Sustainable ecological development options *Global Change effects	Assessment of Industry and Infrastructure *Technological development options *Industrial development options *Infrastructural options	Market Assessment *Identification of market possibilities *Possible market strategies	Socio-economic Assessment *Socio-economic development options
Block III Integrated Analyses	Integrated analyses of all components based on results from Block II				
Block IV Policy Implications	*Allowable cut *Forest management	*Strategies for sustainable development	*Industrial strategies *Infrastructural strategies	*Market strategies *Marketing activities	*Socio-economic development strategies

Table 1.
Analytical Structure of the Siberian Forest Study.

A very important sector of the resource part of the project is the prediction of the dynamics of the forest fund — long-term and short-term forecasts. This resource of temporal changes of structure and parameters of the forest ecosystems is carried out as a result of modelling the processes of natural growth and the development of woodstands and complex influence of anthropogenic and natural factors. Formalized descriptions of the forest fund as an object of modelling is based on the functional classification of the type of forest growth conditions, the main forest species age-group classes and productivity classes. The changes of structure and state of forest fund is described over time as a result of the forest ecosystems natural succession and anthropogenic and natural influences. The forecast of dynamics of the forest fund is given as a number of possible trajectories of forest stands development corresponding to different strategies of forest utilization and reproduction of forest resources and level of forest protection from pests and diseases.

The predictive estimations of dynamics of the forest fund are used for the grounding of acceptable scales of inexhaustible forest utilization by forest management enterprises and regions of Siberia, agreed volumes of forest utilization and demand for wood raw material, optimal placement of assortment plans for felling for forest enterprises.The latter is carried out as a result of solving mathematical programming problems, the functional of which is the maximum income from felling and demand of wood is used as a restriction. The development of a scheme of the transport from territories must be a preliminary stage of solving this task. The selection of acceptable strategies and formation of regional programs of managing and reproduction of forest resources must be based on ecological and economic priorities. Generation of the data base and mathematical apparatus for solving this task for concrete territories and regions is the most important part of the implementation of the project.

2. Biospheric Role of the Forest of Siberia, Their Interaction with Global Changes

The goal of this subproject is to give a quantitative assessment of ecological and social functions of the forest ecosystems and to forecast their conditions taking into account the recommended strategies of socio-economic development and scenarios of global changes by the end of the twenty first century.

The goals of the subproject include:

- assessment of ecological conditions of the main types of forest ecosystems of Siberia within basic territorial units (ecoregions);
- definition of interaction parameters between forest ecosystems and gas composition of the atmosphere;
- modelling the dynamics of forest cover under conditions of global climatic changes;
- working out a strategy of preserving biodiversity and its utilization in the systems of complex nature use.

On the basis of a data base on forest resources and specialized information on their conditions qualitative characteristics are identified of the ecological functions of Siberian forests on the global and regional levels.

For the development of long-term scenarios of inexhaustible utilizations of forest and retaining the ecological significance of forests models have been elaborated with optimal structures of forest ecosystems corresponding to the criteria of a steady development of the biosphere. Special attention is paid to the process of transfer substance and energy and their effect on the atmosphere gases. On the basis of the assessment of carbon pool in terrestrial phytomass, soils and marshes models models of carbon fluxes have been created which reflect natural and anthropogenic processes occurring in forests (forest management activities, forest fires *et al*).

A very important scientific problem dealing with antropogenic effects on biosphere is a forecast of the way vegetation cover responds to climatic changes.One of the methods of solving the problem is the establishment and analysis of mathematical models of dynamics of vegetation zones. At present while creating such models evaluations of vegetation cover distribution by bioclimatic schemes for equal climatic scenarios are being widely used which are obtained from the general circulation of the atmosphere. The project proposes a new approach based on the investigation of dynamics in the shift of vegetation zone boundaries and unequal climatic scenarios which drastically improve the forecast assessments. In the framework of this steady analysis of the parameters of the global carbon cycle a new approach to imitation modelling and assessment of carbon pools has been worked out which is based on the model of dynamics in the shift of bioms boundaries in Siberia under the pressure of long-term and short-term process.

The elaboration of parameters for ecological assessment of forest ecosystems must include the problems of biodiversity. This is particularly important in choosing the ways of forest management and forest industry activities. It is necessary to work out a complex of requirements for the system by which nature is used which secures the preservation of biodiversity applied to the conditions of Siberia and gives a predicting assessment of this problem by the end of the XXI century despite a variety of management strategies and global changes.

3. Geoinformation System "Forest Resources of Siberia (GIS FRS)"

A very important stage in the implementation of the Siberian project is to design and activate the Geoinformation system which enables an operative solution on different levels in forest management — federal, regional and local. GIS FRS is being designed as a computerized system of collecting, storing and actualizing factographic and cartographic data on the forest resources and natural-economic features of Siberia; the system of analysis and inventory of these data for the purpose of assessing the ecological potential of Siberian forests and their role in promoting a sustainable socio-economic development of the country.

The functional structure of GIS FRS is determined by the goals solved in the framework of three major subprojects (resource, ecology and economy) and by the necessity of providing domestic and international public organizations with information on the current state (condition) of boreal forests. The list of goals being solved in the framework of resource, ecological and economic blocks of the project is given in the technical proposals on the elaboration of the relevant projects.

GIS FRS includes the system of data classification, interpretation and data base. The system of classification and interpretation is worked out in accordance with international and domestic systems by taking into account the structure of the data base GIS.

The composition and structure of technical software of GIS FRS are determined by a necessity of forming, storing and processing a huge (up to 10 gbite) factographic and cartographic data base and solving a complex of scientific-technical problems on the main divisions of the project, along with the necessity to incorporate GIS into the international computer network.

The perspective of GIS FRS development is closely correlated with the development of Siberian Forest monitoring which is being carried out on the basis of a special project "TAIGA" done cooperatively by Russia and the USA by a joint Russian-American working group "Earth Sciences" in accordance with the decision of the Joint Commission Chernomyrdin-Gore on Energy and Exploration of Outer Space. AVHRR data from polar satellites NOAA, ground data, aerial survey data are supposed to be used in the joint project. The Installation of station HRPT NASA is planned in the Siberian International Centre for Ecological Research of Boreal Forests (Krasnoyarsk) for getting satellite data.

The realization of the project will be done in the following directions:

1. Detection and mapping of the active forest fires by using AVHRR data. A task is set to work out and use methods of recording spatial and temporal distribution of an active forest fire on the basis of detection of thermal anomalies by sensor AVHRR. The daily monitoring of active fires will give information on the spread broadcast of spots of inflammation and effectiveness of their localization. AVHRR data can also be used for inventory of burned areas, intensity of damaged forest stands and assessment of reforestation processes on the burns. The obtained information is included into the system of federal, regional and local GIS forest fires monitoring of Russia and used for the purpose of optimizing forest fire protection.
2. Assessment of bioproductivity using AVHRR is regarded as an important task of monitoring the forests of Siberia. The expert estimations indicated a wide range in the framework of GIS LS of perspectives for mapping spatial distribution of biomass in the boreal forests and marginal areas, assessment of dynamics of forest vegetation and work out models of productivity of forest ecosystems.

Cartographic and factographic data on ecological, meteorological and vegetation characteristics of land utilization and state of forest cover will be used in the GIS structure. The analysis of this data will enable us to assess the effect of demolishing factors (fires, damage by insect pests, felling etc.) and restrict them.

Forest monitoring by the application of remote sensing envisages a wide spectrum of usage data from the Russian, American and European satellites for the purpose of organizing and maintaining an effective system of surveying the forest over of Siberia. The "TAIGA" project is supposed to be a major stage in solving this problem.

♦

GLOBAL CHANGE, THE GLOBAL BOREAL FOREST, AND FIRE: SEARCH FOR NEW STRATEGIES IN SCIENCE POLICIES AND RESEARCH MECHANISMS

JOHANN GEORG GOLDAMMER
Max Planck Institute for Chemistry
Biogeochemistry Department
Fire Ecology Research Group
c/o Freiburg University
D-79085 Freiburg
Germany

and

VALENTIN V. FURIAEV
Sukachev Institute for Forest and Wood
Russian Academy of Sciences, Siberian Branch
Akademgorodok
RU-Krasnoyarsk 660036
Russian Federation

ABSTRACT. One of the first priority areas among joint East/West research programmes is th rational use of natural resources and sustainable development of regions. The global boreal fores is one of the most productive but also highly vulnerable ecosystems. Because of its size an ecological functions the boreal forest and its most dynamic disturbance factor fire play a important role in ecosystem processes at a global scale. New research initiatives between Wester and Eastern countries are designed to address a series of phenomena and problems. Cooperativ research agreements under the International Geosphere-Biosphere Programme (IGBP), i conjunction with the International Boreal Forest Research Association (IBFRA), have been use as research instruments to initiate a joint research campaign. The Fire Research Campaign Asia North (FIRESCAN) was implemented under these agreements. This contribution highlights th history of this research and gives recommendations for future use of the successfully applie research mechanisms.

1. Introduction

The circumpolar boreal forest biome occupies nearly one third of the total global forest area More than seventy percent of the global boreal forest is in Eurasia, mainly in the Russia Federation; the remainder covers large parts of Canada and Alaska. The ecological importanc

V A Koptyug and J Klerkx (eds), Science Policy New Mechanisms for Scientific Collaboration between East and West, 45–61

and influences of the boreal forest on the zonal and global atmosphere and climate is less understood than its economic importance. However, the magnitude of the boreal forest area suggests that it plays an important role in ecosystem processes on a global scale. Large-scale disturbance and destruction of the boreal forest due to human activities may contribute to regional and global climate change. Vice-versa, global climate change, induced by the cumulative effects of human activities and natural environmental changes, may have considerable influence on the future of the circumpolar boreal forest.

This contribution intends to highlight the role of natural and human-caused fires in relationship to ecology, resource protection needs, economy and global ecological processes. Since the global boreal forest area is mainly located in territories belonging to the Russian Federation, northern European countries, and North America, the research objectives need to be addressed jointly through East-West collaboration. Furthermore, the complexity of envisaged research requires science policy instruments which are open to interdisciplinary approaches, flexible responses, and sustainable programmes.

Experience collected through international research programmes during the past years reveals that these requirements can be met with objective- or task-oriented, rather short-term living projects, limited to a couple of years. However, in the case of East-West projects one needs to take into account the different evolution of science and science instruments during the historical years of separation and isolation. The prerequisites for cooperation between the future partners are completely different. Most striking are the consequences of restructuring the national economies in the Eastern European countries and the CIS. Thus, the collaborative research and partnerships must be developed accordingly.

2. The Boreal Forest: Extent and Economic Importance

The world's total boreal forests and other wooded land within the boreal zone cover 1.2×10^9 ha of which 920×10^6 ha are closed forest. The latter number corresponds to ca. 29% of the world's total forest area and to 73% of its coniferous forest area (ECE/FAO, 1985). About 800×10^6 ha of boreal forests with a total growing stock (over bark) of ca. 95 billion m^3 are exploitable (41% and 45% respectiveley of the world total). The export value of forest products from boreal forests is ca. 47% of the world total (Kuusela, 1990, 1992).

The vast majority of the boreal forest lands (taiga) of Eurasia are included in the Russian Forest Fund, covering ca. 900×10^6 ha. Depending on the criteria used to define "boreal forest", the area of closed boreal forest in the Russian Federation varies from 400 to 600×10^6 ha (Pisarenko and Strakhov, 1993). These numbers correspond to a 43-65% share of the world's closed boreal forest.

Natural fire (lightning fire) is an important ecological factor in the boreal forest system. Under given climate and site conditions, fire is the most important abiotic factor controlling age structure, species composition and physiognomy, shaping landscape diversity and mosaic, and influencing energy flows and biogeochemical cycles, particularly the global carbon cycle (cf. monographs and synopses e.g. by Sofronov, 1967; Slaughter et al., 1971; Zackrisson, 1977; Sherbakov, 1979; Viereck and Schandelmeier, 1980; Alexander and Euler, 1981; Heinselman, 1981; Wein and MacLean, 1983; Kurbatsky, 1985; Johnson, 1992; Sannikov, 1992; Furiaev, 1994; Goldammer and Furiaev, 1994). In Eurasia fire has long been an important tool for land clearing (conversion of boreal forest), silviculture (site preparation and improvement, species selection) and in maintaining agricultural systems, e.g. hunting societies, swidden agriculture, and pastoralism (Viro, 1969; Pyne, 1994). In addition to the natural fires, these old cultural practices brought a tremendous amount of fire into the boreal landscapes of Eurasia. In the early 20th century, fire-use intensity in the agricultural sector began to decrease because most of the deforestation had been accomplished, and traditional small-sized fire systems (treatment of vegetation by free burning) became replaced by mechanized systems (use of fossil-fuel burning for mechanical treatment). Despite the loss of traditional burning practices, however, humans are still the major source of wildland fires; only 15% of the recorded fires in the Russian Federation are caused by lightning (Korovin, 1994; Fig.1).

The elimination of wildfires in Western Eurasia (Norway, Sweden, Finland) during recent years (Stocks, 1991; Kuusela, 1992) leaves Eurasia's major fire load on the territory of the Russian Federation and other countries of the Commonwealth of Independant States (Tab.1). Statistics compiled by the Russian Aerial Fire Protection Service *Avialesookhrana* show that between 10,000 and more than 30,000 forest fires occur each year, affecting up to 2-3x10^6 ha of forest and other land (Fig.2). Since fires are monitored (and controlled) only on protected forest and pasture lands, it is estimated that the real fire load in the whole of Eurasia's boreal vegetation is much higher, probably reaching up to 10x10^6 ha in extreme fire years.

Tab.1. Selected basic data on global and Eurasian boreal forest fires

Boreal Forest/Wildland Fires	
Boreal North America (Annual Average)	1-5 x 10^6 ha
Extreme Years, e.g. Canada 1989	7.4 x 10^6 ha
Boreal Western Europe (Annual Average)	< 4,000 ha
Extreme Years, e.g. Sweden 1933	30,000 ha
Boreal China (Annual Average)	< 55,000 ha
Extreme years, e.g. NE China 1987	1.3 x 10^6 ha
Boreal Eurasia, Recorded Fires (Annual Average)	2-3 x 10^6 ha
Extreme Years, e.g. 1987	?
Boreal Eurasia, Unrecorded Fires (on Non-Protected Lands)	10 x 10^6 ha?

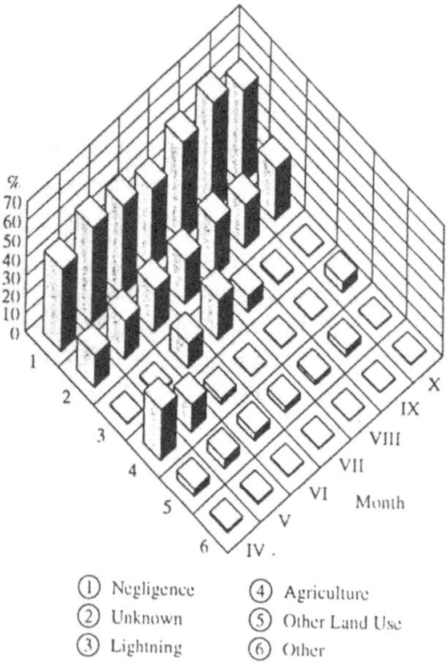

Fig.1. Monthly distribution of causes of wildfires on protected lands in the USSR and the Russian Federation between 1947 and 1992 Source· Avialesookhrana (Korovin, 1994).

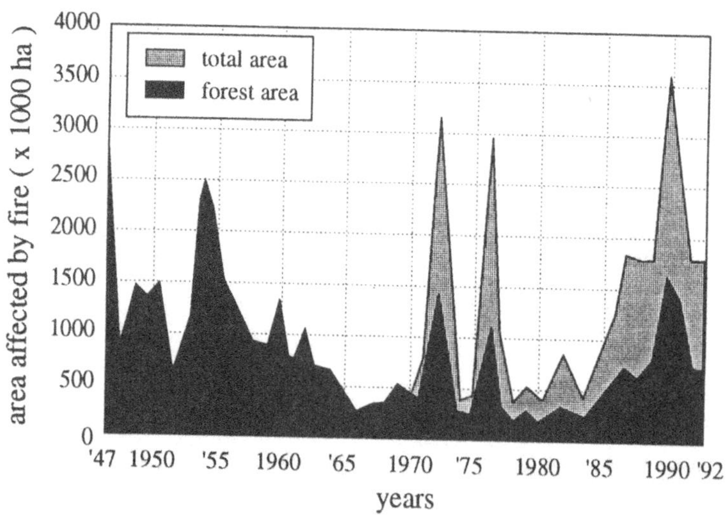

Fig.2. Area of protected lands affected by wildfires in the USSR and the Russian Federation between 1947 and 1992 Source· Avialesookhrana (Korovin, 1994).

3. Global Change, the Boreal Forest, and Fire

Expected global warming over the next 30-50 years, as predicted by Global Circulation Models (GCM's), will be most evident in the northern circumpolar regions (Bolin et al., 1986; Maxwell, 1992; Shugart and Smith, 1992; Shugart et al., 1992). As Wein and de Groot (1993), Stocks (1993) and Stocks et al. (1994) underscore, fire may be the most important (widespread) driving force in changing the taiga under climatic warming conditions. The prediction of increasing occurrence of extreme droughts in a $2xCO_2$ climate indicates that fire regimes will undergo considerable changes. An increase in the length of the fire season will lead to a higher occurrence of large, high-intensity wildfires. Such fire scenarios may be restricted to a transition period until a new climate-vegetation-fire equilibrium has been established.

Regional warming will also lead to the shift of vegetation zones, e.g. causing the boreal forest to shift north ca. 500-1000 km (Kauppi and Posch, 1988). The shift of ecosystems will have a considerable impact on the distribution of phytomass. Estimates of carbon stored in living and dead plant biomass (without soil organic matter) above- and belowground in the global boreal forest area range between 66 and 98 Gt (66-98x10^{15} g) (US Department of Energy, 1983; Apps et al., 1993). Additional large amounts of carbon are stored in the boreal forest soils (ca. 200x10^{15} g) and in the boreal peatlands (ca. 420x10^{15} g) (Apps et al., 1993; Tab.2). It is feared that changes in fire regimes due to climate change will affect the balance of the boreal carbon pool and lead to the additional release of carbon into the atmosphere, thus acting as temporary feedback loop to global warming.

Changing forestry practices in boreal Eurasia, stimulated by increasing national and international demands for boreal forest products, have resulted in the widespread use of heavy machinery, large-scale clearcuts, and thereby in the alteration of the fuel complexes. The opening of formerly closed remote forests by roads and the subsequent human interferences bring new ignition risks. Additional fire hazards and environmental consequences which are still mainly unpredictable are created on forest lands heavily damaged by industrial emissions (severe damages in the Russian Federation are on ca. 9x10^6 ha). Radioactive contamination on an area of ca. 7x10^6 ha creates considerable problems in redistribution of radionuclides through forest fires (Dusha-Gudym, 1992, 1994; Fig.3). These direct effects on the ecosystem are added to the indirect effects of climate change, and both will certainly lead to an unprecedented era of fire.

4. East-West Research in Transition: Integration of Research Structures and Mechanisms

As in many other disciplines of science, the historical development of boreal environmental research, and specifically forest fire research, clearly suffered from being separated for at least four to five decades by political boundaries between East and West. Except for some courtesy contacts between fire managers and scientists during the Cold War and some occasional joint participation at international meetings, no exchange or cooperation in fire science took place until the very recent of political isolation. Under the given and predicted developments of the global boreal forest, it is evident that new initiatives in cooperative fire research are needed. The differences in research structures and economic situations in the East and the West, however, will require specific approaches to perform joint research programmes.

Tab.2. Contemporary boreal forest biome areas and C pools. Totals may not agree due to rounding errors. For more details on sources of investigations see Apps et al. (1993).

	Area (10^6 ha)		Carbon Pools (Pg C)					
	Area	Peat Land	Plant Biomass	Plant Detritus	Forest Soil	Peat	Forest Products	Total
Alaska	52	11	2	1	10	17	<0,1	30
Canada (Boreal Forest Biome)	304	89	8	*	65	113	0,2	186
Canada (Cordilleran)	72	3	6	*	16	4	0,3	27
Russian Federation	760	136	46	31	100	272	2,9	451
Scandinavia	61	20	2			13		15
Total	1249	260	64	32	199	419	3,4	709

* Plant detritus estimates are included in soil C pool.

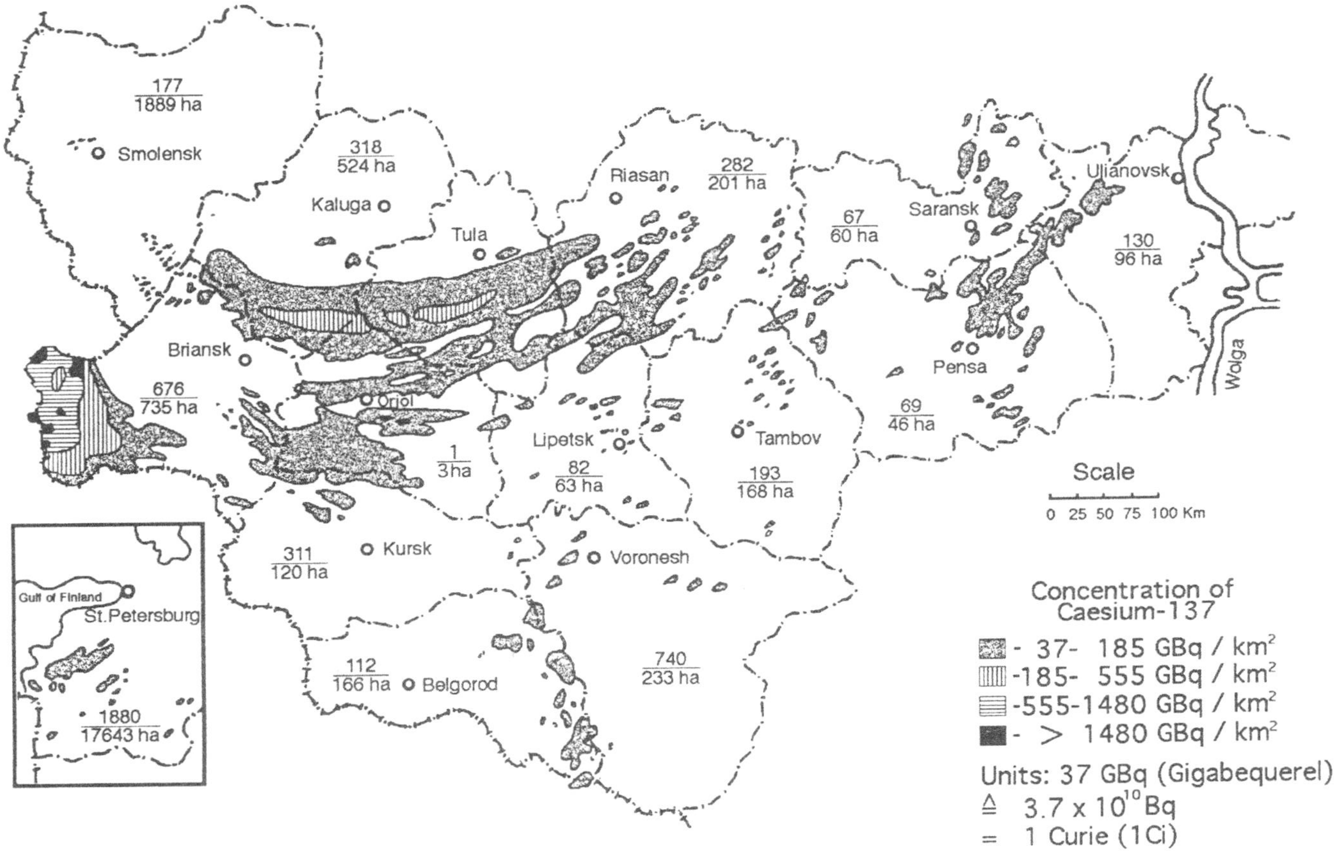

Fig.3. Land area in the Russian Federation contaminated by Caesium-137 (main map shows territory East of Chernobyl nuclear power plant). The numbers within the districts (Oblasts) give the number of forest fires and the total area affected by fire in 1992 (from Dusha-Gudym, 1994).

The research facilities of the former USSR, capable of supporting regional fire research programmes, are mainly part of the Academy of Sciences. The major research centers are now located in the Russian Federation. These rather large research complexes are suffering heavily under Russia's current economic problems. At present, the total funding made available to the Academy of Sciences is so insufficient that adequate salaries cannot be paid. In most places there is no funding available for operating costs, investment or maintenance of research hardware and infrastructures.

At the same time, many institutes belonging to the Academy are over-staffed. Taking into account that in many places virtually no funding for operating research is left, the science community is restricted to theoretical work.

In the Western countries the present economic depression, coupled with high unemployment rates, has severe repercussions on research. Increasing costs for advanced research hardware coincide with decreasing research budgets. However, at the same time the field of funding mechanisms is becoming more diverse. New instruments are created by intergovernmental research agreements such as the International Geosphere-Biosphere Programme (IGBP) and the International Boreal Forest Research Association (IBFRA), or private research foundations. They open numerous ways to obtain funding for short-term projects, thus creating enormous opportunities. In the context of environmental sciences, particularly in forest- and global-change related research, funding instruments of interest are, among other, the International Science Foundation (Washington, USA), the European Forestry Institute (Joensuu, Finland), or even the recently founded International Center for Science and Technology (Moscow, Russian Federation).

The prerequisites for using these research instruments are the availability of permanently operating research nodes, e.g. the Russian Academy of Sciences in the East and universities or government-funded laboratories in the West. Only these are able to manage short-term objective- or task-oriented research programmes.

The first phase of East-West programmes in forest and fire research is now underway. Its outline, objectives and implications for further activities are described in the following.

5. Initiation of Cooperative East-West Forest Fire Research in Boreal Eurasia

In 1990-91 first initiatives were launched towards cooperative fire research. At the invitation of the Aerial Forest Fire Protection Service Avialesookhrana and the Academy of Sciences of the USSR, S.J.Pyne (Arizona State University, Phoenix, U.S.A.) and J.G.Goldammer (Max Planck Institute for Chemistry, Germany) explored the most important fire research and management facilities and programs of the Soviet Union (Pyne, 1991, 1992). One of the outcomes was the decision to jointly prepare and conduct a conference and a field campaign, both devoted to the exploration of fire in ecosystems of boreal Eurasia.

The objective of the conference, which was held in Krasnoyarsk, 29 June to 3 July 1993, was to compile, interpret, and dicuss the state of knowledge on the role and impacts of fire in boreal ecosystems, with special emphasis on Eurasia. The publication of the re-examined knowledge aims to stimulate a new era of pan-boreal fire research, especially considering the need to put basic and specifical local aspects of fire ecology into the broader context of the newly emerging global fire science (Crutzen and Goldammer, 1993; Goldammer, 1994). The monograph "Fire in Ecosystems of Boreal Eurasia" is in preparation at this stage (Goldammer and Furiaev, 1994).

The re-examination of the state of knowledge in boreal fire science was followed by the first joint fire experiment. The Fire Research Campaign Asia-North (FIRESCAN) was prepared and implemented in its first phase under the co-sponsorship of an IGBP subprogramme and IBFRA. The research campaign was largely financed through support from the Volkswagen Foundation.

The International Geosphere-Biosphere Programme (IGBP) provides a base for interdisciplinary research programs. One of the operational IGBP core projects is the International Global Atmospheric Chemistry Project (IGAC). One of the activities of its foci (Focus 2: Natural Variability and Anthropgenic Perturbations of the Tropical Atmospheric Chemistry) is oriented towards investigating the impact of biomass burning on the atmosphere and biosphere ("Biomass Burnig Experiment" [BIBEX])(IGAC, 1992).

The boreal fire research programme is a joint activity with the International Boreal Forest Research Association (IBFRA). Following the recommendations of the "White Sea Declaration" of 1990, IBFRA was founded in 1991 in Mezhgorje, Ukraine (Fosberg, 1992). The two priority research areas are (1) Inventory, Monitoring, and Classification of Boreal Forests and (2) Global Climate Change and Ecosystem Function of Boreal Forests (Fig.4). Under the second area the first IBFRA Working Group established was on Stand Replacement Fire. In the context of the eight Working Hypotheses developed by the group in 1992, the fire experiment scheduled for 1993 was considered an IBFRA activity.

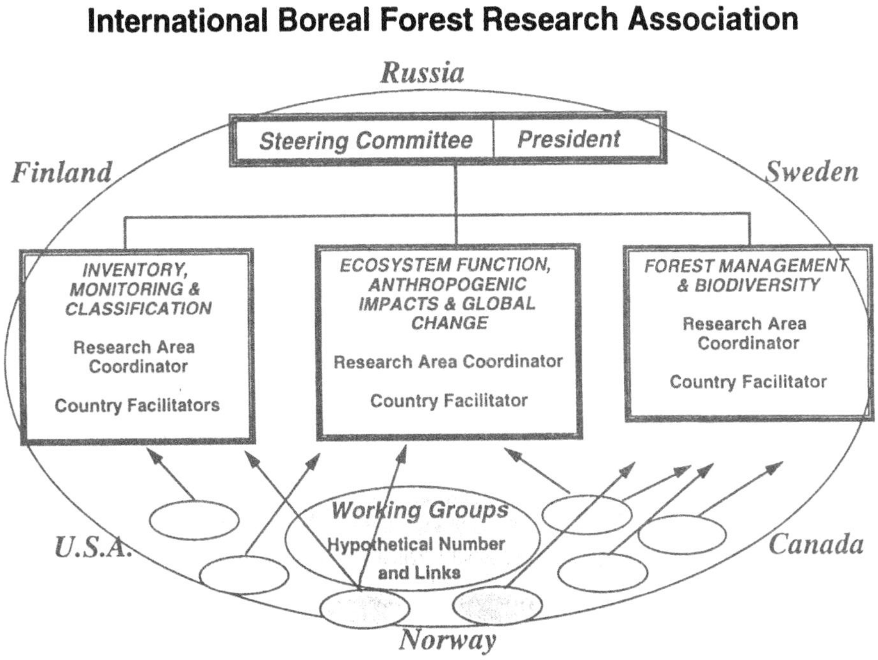

Fig.4. Organizational diagram of the International Boreal Forest Research Association (IBFRA).

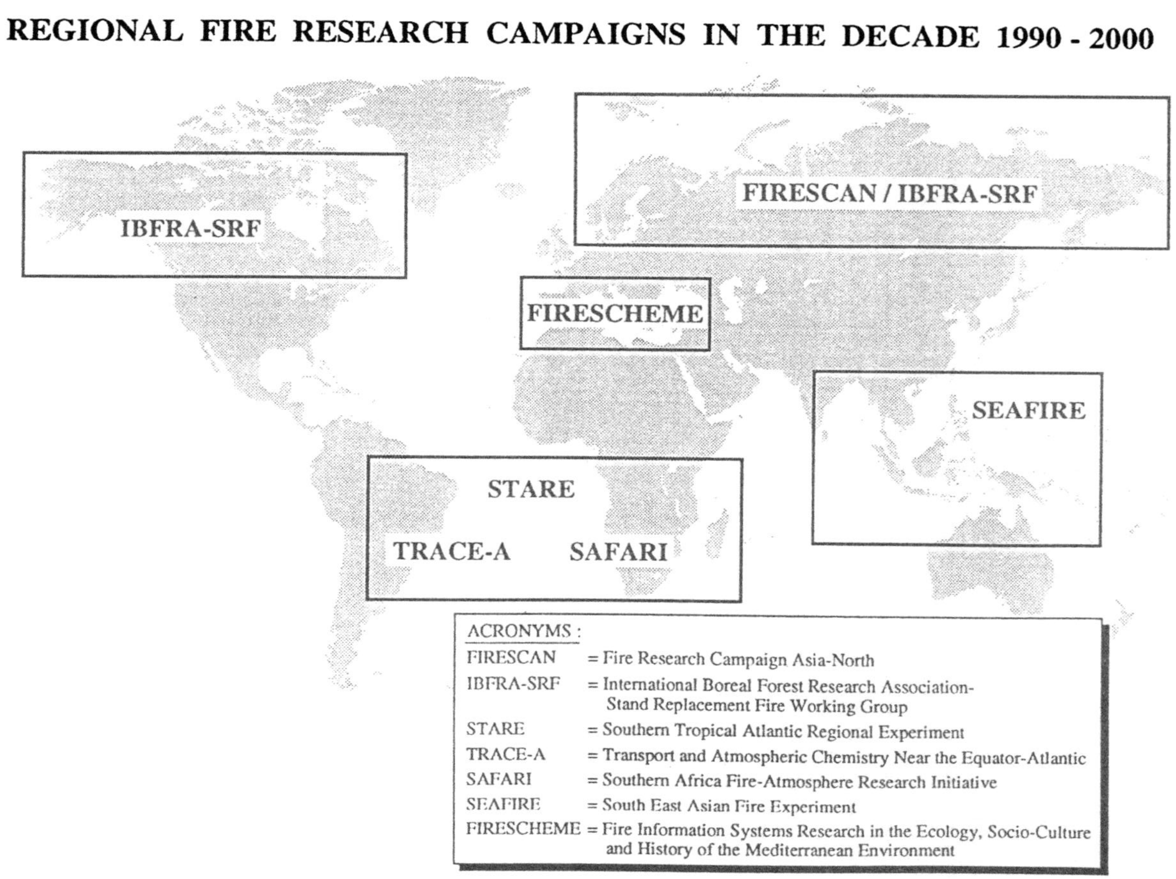

Fig.5. World map showing the target areas of regional fire research campaigns which are in the stage of implemementation or planning (from Goldammer, 1994).

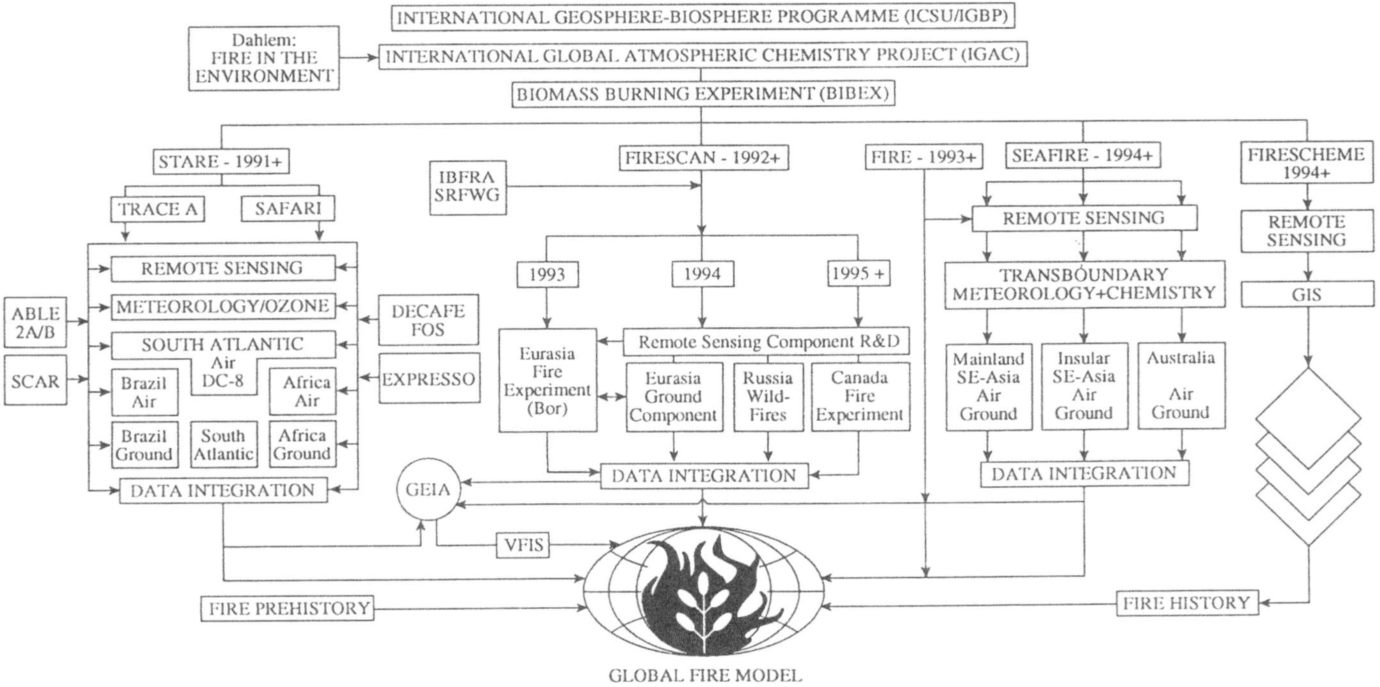

Fig.6. Organizational diagram of fire research activities which will be integrated into the development of a global fire model. Acronyms are explained in the main text (Goldammer, unpubl., 1994).

List of Acronyms of Fire Research Activities shown in Figure 6

BIBEX	Biomass Burning Experiment (→ IGAC)
EXPRESSO	Experiment for Regional Sources and Sinks of Oxidants
FIRE	Fire in Global Resource and Environmental Monitoring (CEC-JRC)
FIRESCAN	Fire Research Campaign Asia - North (→ IGAC)
FOS/DECAFE	Fire of Savannas/Dynamique et Chimie Atmosphérique en Forêt Équatoriale
GEIA	Global Emissions Inventory Activity (→ IGAC)
IBFRA	International Boreal Forest Research Association
ICSU	International Council of Scientific Unions
IGAC	International Global Atmospheric Chemistry Project (→ IGBP)
IGBP	International Geosphere-Biosphere Programme (→ ICSU)
SAFARI	Southern African Fire-Atmosphere Research Initiative (→ IGAC)
SEAFIRE	South East Asian Fire Experiment (→ IGAC)
STARE	Southern Tropical Atlantic Regional Experiment (→ IGAC)
SRFWG	Stand Replacement Fire Working Group (→ IBFRA)
TRACE-A	Transport and Atmospheric Chemistry Near the Equator - Atlantic
VFIS	Vegetation Fire Information System (→ Dahlem Konferenz Model)

The various regional fire research campaigns organized under the umbrella of IGBP/IGAC/BIBEX are operational or in the planning stage (Fig.5 and 6). The campaigns are designed to establish a comprehensive analysis of environmental impacts of fire on a global scale. The East European/North Asian part of this global research approach, as represented by the regional FIRESCAN programme, is a contribution in which East-West collaboration is mandatory.

The first FIRESCAN activity mentioned above was a large experimental forest fire in Central Siberia (Fig.7). In this fire experiment the ecological and atmospheric chemical impacts of a high-intensity fire were investigated by a mixed group of scientists from all boreal countries (FIRESCAN Science Team, 1994a). This research had value beyond the symbolic sharing of East-West expertise in fire research. The results have already brought new insights into the details of fire history, ecology, and behaviour. Some of the emission characteristics observed were new to fire research, e.g. the emission of high amounts of radiatively active trace gases and compounds affecting the stratospheric ozone layer (cf. also Manö and Andreae, 1994).

In addition, a long-term follow-up research area has now been established, the first one in the whole of Eurasia's taiga forests. This site will serve as a research focus for the next decades and centuries. For details cf. data published in various journals and the monograph in prep (FIRESCAN Science Team, 1994a,b,c; Goldammer and Furiaev, 1994).

In November 1993 the "Taiga Aerospace Investigations using GIS Applications (TAIGA) Workshop" was held in Moscow (Stocks 1994). The workshop, sponsored by NASA's Office of Mission to Planet Earth and the International Forestry Institute (IFI) of the Russian Academy of Sciences, was aimed to develop a multi-national science plan for studying the boreal forests of Russia. The impetus for this workshop was the decision to place new satellite receiving stations in Siberia, a region of the world where previous satellite coverage was less than adequate, and to use this increased capability to intensively study the vast boreal forests of Russia to determine their role in global biogeochemical cycles. The first NASA HRPT (High Resolution Picture Transmission) station will be installed in Krasnoyarsk in 1994. Cooperative projects developed at the workshop will make use of both AVHRR (Advanced Very High Resolution Radiometer) data from NOAA polar-orbiting satellites and existing ground and aircraft based datasets.

Three subgroups (Forest Fire, Productivity and Carbon Budget, and Mathematical Modelling) were created at the workshop for the purpose of focussing discussion. The Forest Fire Subgroup proposed the following studies, all of which require the extensive use of satellite imagery and Russian forest inventory and other ground truth data:

- Detection and mapping of active fires using AVHRR imagery.
- Monitoring the spatial distribution of live fuel moisture using AVHRR imagery and ground truth data.
- Determine the spatial and temporal distribution of ground layer fuel moisture, using satellite and ground station data, and integrate this information using a multi-index fire danger/behaviour system (the Canadian Forest Fire Danger Rating System will be evaluated in this study).
- Evaluation of lightning risk using AVHRR imagery to track and assess thunderstorms.
- Investigate the relationship between the atmospheric boundary layer and intense forest fire behavior.
- Develop an annual inventory and assessment of the area burned by fires in all regions of boreal Russia.

Many of these collaborative Russian/Canadian/United States forest fire research initiatives complement or reinforce a number of studies already planned by the Stand Replacement Fire Working Group of IBFRA and FIRESCAN respectively.

6. Lessons Learned

After the initial phase of FIRESCAN we found that the capabilities as well as the mutual will to conduct this joint research campaign have been successfully overcome shortcomings in research funding and the infrastructural and organizational limitations. The mechanisms of international research agreements have assembled and coordinated highly qualified research individuals and groups from various disciplines and countries of origin. They were able to bring in their specific expertise and funding, and, more importantly, enthusiasm. The spirit of the joint research campaign is reflected by the procedure of publishing research results as papers jointly written by the FIRESCAN Science Team, and not by individuals (cf. FIRESCAN Science Team, 1994a,b,c).

The follow-up of the 1993 campaign is not restricted to the immediate research targets addressed and the individuals who participated in the first phase. From the beginning of the campaign new follow-up research agreements multiplied instantaneously.

The explanation why this research campaign had been planned and implemented within a relatively short time lies in the nature of the research mechanisms within the frame of IBGB/IGAC/BIBEX and IBFRA. None of these multilateral agreements on the working group or "field" level involves any large bureaucracy or centralized management of the various research components. The only "central" function is the coordination of the campaign. This coordination is done through individual researchers on various levels. Since the researchers themselves have limited or no secretarial support, all administrative or coordinative actions are restricted to an absolute minimum.

Advanced communication systems, however, are an important prerequisite for the planning and evaluation phases. The availability of electronic mail in the Russian Federation has bridged the gap of communication links wherever it was difficult to directly access places through telex or telephone/facsimile systems. The envisaged set up of an East-West scientific DATA NET certainly will improve information flow and joint work on data gained through campaigns such as FIRESCAN

Fig.7. Aerial view of the Bor Forest Island Fire Experiment of July 1993 The results of this first East-West fire experiment have given important insights into the ecology and atmospheric chemical impacts of boreal forest fires (Photo J G Goldammer).

7. Conclusions: Implications for Future Cooperative Research

The conclusions derived from the experience gained in setting up mechanisms in forest ecosystems and global change-related research between the East and the West, forest fire research being an example for such kind of activities, are encouraging The priorities have been recognized, relevant research programmes defined, and the scientific capabilities and the enthusiasm to implement complex research agendas available

The critical period ahead of us is burdened with the contemporary restructuring of the national economies of the CIS. It must be overcome through solidarity by researchers and research funding mechanisms from the West. As the NATO conference statement (this volume) underscores, priority must be given to keeping the human capital and sciencific potential alive. The research mechanisms described in this paper will provide a contribution.

8. References

Alexander, M. E. and D. L. Euler, Ecological role of fire in the uncut boreal mixedwood forest, in Canadian Forestry Service, Department of the Environment, editor, *Boreal Mixedwood Symposium*, CODJFRC Symp.Proc. O-P-9, 42-64 (1981).

Apps, M. J., W. A. Kurz, R. J. Luxmore, L. O. Nilssohn, R. A. Sedjo, R. Schmidt, L. G. Simpson, and T. S. Vinson, The changing role of circumpolar boreal forests and tundra in the global carbon cycle. *Water, Air, and Soil Pollution*, 70, 399 (1993).

Bolin, G., B. R. Döös, J. Jäger and R. A. Warrick, *The greenhouse effect, climate change and ecosystems*. SCOPE 29. J. Wiley and Sons, Chichester, England (1986).

Crutzen, P. J. and J. G. Goldammer (eds.), *Fire in the Environment. The Ecological, Atmospheric Chemical, and Climatic Importance of Vegetation Fires*. Environmental Sciences Report 13. John Wiley and Sons, Chichester, England (1993).

Dusha-Gudym, S. I., Forest fires on areas contaminated by radionuclides from the Chernobyl power plant accident. *Int. Forest Fire News* (ECE/FAO) **No.7 (August 1992)**, 4 (1992).

Dusha-Gudym, S.I., The effect of forest fires on radionuclide concentrations and transfer of radioactive contamination of forest fuels and their combustion products, in J. G. Goldammer and V. V. Furiaev, editors, *Fire in Ecosystems of Boreal Eurasia* (in prep., ca.1994).

ECE/FAO (Economy Commission for Europe/Food and Agricultural Organization of the United Nations), *The Forest Resources of the ECE Region (Europe, the USSR, North America)*. ECE/FAO/27, Geneva (1985).

FIRESCAN Science Team (J.G.Goldammer, coord.), Fire in boreal ecosystems of Eurasia: First results of the Bor Forest Island Fire Experiment, Fire Research Campaign Asia--North (FIRESCAN). *J. World Resources Rev.* (1994a, in press)

FIRESCAN Science Team, The Bor Forest Island Fire Experiment (I): Fire history and ecology of a boreal pine forest in the central taiga region of the Dubches plain, Western Siberia, *J. Wildland Fire* (1994b, in prep.).

FIRESCAN Science Team, The Bor Forest Island Fire Experiment (II): Ecological and atmospheric chemical effects of a high-intensity stand replacement fire in a boreal pine forest in the central taiga region of the Dubches plain, Western Siberia, *J. Wildland Fire* (1994c, in prep.).

Fosberg, M., International Boreal Forest Research Association, Stand Replacement Fire Working Group. *Int. Forest Fire News* (ECE/FAO) **No.7 (August 1992)**, 6 (1992).

Furiaev, V. V., *The Role of Fire in Forest Formation Processes*, Int. Ass. Wildland Fire (in press)

Goldammer, J. G., Interdisciplinary research projects for developing a global fire science, *12th Conference on Fire and Forest Meteorology*, 26-28 October 1993, Jekyll Island, GA (1994, in press).

Goldammer, J. G. and V. V. Furiaev (eds.), *Fire in Ecosystems of Boreal Eurasia*. N.N. Publ. (in prep., ca.1994).

Heinselman, M. L., Fire intensity and frequency as factors in the distribution and structure of northern ecosystems, in H. A. Mooney et al., editors, *Fire Regimes and Ecosystem Properties*,7, USDA For. Ser. Gen. Tech. Rep. WO-26 (1981).

IGAC (International Global Atmospheric Chemistry Project), *Biomass Burning Experiment: Impact on the Atmosphere and Biosphere. An Activity of the International Global Atmospheric Chemistry (IGAC) Project*, IGAC Core Project Office, The MIT, Cambridge, MA (1992).

Johnson, E. A., *Fire and Vegetation Dynamics. Studies from the North American Boreal Forest.* Cambridge University Press, Cambridge, UK (1992).

Kauppi, P. and M. Posch, A case study of the effects of CO_2-induced climatic warming on forest growth and the forest sector: A. Productivity reactions of northern boreal forests, in *The Impact of Climatic Variations on Agriculture, Vol 1, Assessments in Cool Temperature and Cold Regions*, 183, Reidel, Dordrecht, The Netherlands (1988).

Korovin, G. N., Statistics on characteristics and spatial and temporal distribution of forest fires in the Russian Federation, in J. G. Goldammer and V. V. Furiaev, editors, *Fire in Ecosystems of Boreal Eurasia* (in prep., ca.1994).

Kurbatsky, N.P. (ed.), *Forest Fires and their Consequences*, Sukachev Institute of Forest and Wood, Krasnoyarsk (1985).

Kuusela, K., *The Dynamics of Boreal Coniferous Forests*, The Finnish National Fund for Research and Development (SITRA), Helsinki, Finland (1990).

Kuusela, K., Boreal forestry in Finland: a fire ecology without fire, *Unasylva* **43 (170)**, 22 (1992)

Manö, S. and M. O. Andreae, Emission of methyl bromide from biomass burning, *Science* **263**, 1255 (1994).

Maxwell, B., Arctic climate: potential for change under global warming, in F. S. Chapin, R. L. Jefferies, J. F. Reynolds, G. R. Shaver and J. Svoboda, editors, *Arctic Ecosystems in a Changing Climate*, 11, Academic Press, New York (1992).

Pisarenko, A. I. and V. V. Strakhov, Global role of the Russian boreal forests: A viable assessment, *World Resources Rev.* (in press).

Pyne, S. J., The Russian fire establishment. Impressions from a study tour. *Int. Forest Fire News* (ECE/FAO) **No.6 (Jan. 1992)**, 3 (1992).

Pyne, S. J., Wildland fire protection in the USSR. report of a study tour, 15 July to 10 August 1991, at the invitation of the Air forest Fire Protection Service of the USSR, conducted by S.J.Pyne and J.G.Goldammer (mimeo, available through the author)(1991).

Pyne, S. J., Wild hearth A prolegomenon to the cultural fire history of northern Eurasia, in J. G. Goldammer and V. V. Furiaev, editors, *Fire in Ecosystems of Boreal Eurasia* (in prep., ca.1994).

Sannikov, S. N., *Ecology and Geography of the Natural Regeneration of Pine Habitats*, Nauka, Moscow (1992).

Sherbakov, I. P. (ed.), *Forest Fires in Yakutia and their Influence on the Nature of Forests*, Nauka (1979).

Shugart, H. H., R. Leemans and G. B. Bonan (eds.), *Boreal Forest Modeling. A Systems Analysis of the Global Boreal Forest*, Cambridge University Press, Cambridge, England (1992).

Shugart, H. H. and T. M. Smith, Modelling boreal forest dynamics in response to environmental change. *Unasylva* **43 (170)**, 30 (1992).

Slaughter, C. W., R. J. Barney and G. M. Hansen (eds.), *Fire in the Northern Environment. A Symposium*, Pacific Northwest Forest and Range Experiment Station, Portland, Oregon (1971).

Sofronov, M. A., *Fires in Forests of Southern Siberia*, Nauka, Moscow (1967).

Stocks, B. J., The extent and impact of forest fires in northern circumpolar countries, in J. S. Levine, editor, *Global Biomass Burning*, 197, The MIT Press, Cambridge, MA (1991).

Stocks, B. J., Global warming and forest fires in Canada, *The Forestry Chronicle* **69**, 290 (1993).

Stocks, B. J., K. C. Ryan and T. J. Lynham, Fire weather climatology across the World's boreal forest zone, in J. G. Goldammer and V. V. Furiaev, editors, *Fire in Ecosystems of Boreal Eurasia* (in prep , ca. 1994).

Stocks, B J., Taiga Aerospace Investigations using GIS Applications (TAIGA) Workshop, *Int. Forest Fire News* (ECE/FAO) **No.10 (January 1994)**, 30 (1994)

US Department of Energy, *Carbon in Live Vegetation of Major World Ecosystems*. DOE/NBB-0037 (1983).

Viereck, L. A. and L. A. Schandelmeier, *Effects of Fire in Alaska and Adjacent Canada - a Literature Review*. BLM-Alaska Tech. Rep. 6 (BLM/AK/TR-80/06), Anchorage (1980).

Viro, P. J., Prescribed burning in forestry, *Comm. Inst. For. Fenn.* **67** (7) (1969).

Wein, R. W. and D. A. MacLean (eds.), *The Role of Fire in Northern Circumpolar Ecosystems*. SCOPE 18, John Wiley & Sons, Chichester, UK (1993).

Wein, R. W. and W. J. de Groot, Fire-climate change hypotheses for the taiga, in J. G. Goldammer and V. V. Furiaev, editors, *Fire in Ecosystems of Boreal Eurasia* (in prep., ca. 1994).

Zackrisson, O., Influence of forest fires on the North Swedish boreal forest, *Oikos*, **29**, 22 (1977).

MAN-MADE CLOSED ECOLOGICAL SYSTEMS: SCIENTIFIC AND APPLIED SIGNIFICANCE, AND PROSPECTS FOR INTERNATIONAL COLLABORATION

J.I. GITELSON, I.I. LISOVSKY and V.E. RYGALOV
International Centre for Closed Ecological Systems,
Institute of Biophysics of SB RAS, Krasnoyarsk, Russia

ABSTRACT. Artificial closed ecological systems (CES) may be used as a tool for experimental investigation of the matter circulation processes and their stability, The creation of space life systems is a strong stimulus for CES studies. Truly functioning CES exist: "Biosphere-2" (USA) and "Bios-3" (Russia). Taking into account the general importance of the problem for mankind, and the high cost of equipment, it is expedient to organize a collective use of installations, with the help of the International CES Research Center, which has the status of an open laboratory. Such an initiative was started in Krasnoyarsk in April 1991. The maximum effect of international collaboration may be achieved through the framework of a long-term experiment on human life support, with an international crew and research team.

Keywords: matter circulation / closed ecological systems / life support / international collaboration.

It is difficult to find a problem more important than the progressive sustainable development of human society in our planet's biosphere.

The infinite existence of the biosphere, taking into account the limited resources of the Earth's biosphere, has been determined by the evolutionary origin and preservation of a closed substance circulation between the living matter and its inanimate surrounding. The drastically increased activity of man, and his intervention into natural cycles of matter circulation have assumed a planetary scale and are becoming a factor which disturbs or even destroys the processes of cyclic conversion of substances on the planet, i.e. the general matter circulation in the biosphere.

It is impossible to confirm or disprove the theoretical estimates of the scale and nature of the consequences of anthropogenous influences on the biosphere by means of direct experiments on it, since they may be dangerous to its existence.

Artificially created, closed ecological systems (CES) of different degree, of complexity may be used as a tool for experimental studies of matter circulation processes and their stability in the Biosphere.

The most informative systems are the CES, which include man as an integral link of the biological substance circulation cycle.

V A Koptyug and J Klerkx (eds), Science Policy New Mechanisms for Scientific Collaboration between East and West, 63–68

A strong stimulus for the creation of CES is the prospective astronautic need for human life support systems, which would be independent from the Earth. Such a system could only be based on closed substance circulation. This problem is quite new for science and technology. It was first set out on in the 50s—60s, in anticipation of Man's imminent space exploration. It was then, that the first experiments on unmanned systems were performed. In 1965, regeneration of man's atmosphere in a closed interchange of gases with photosynthesizing unicellular algae was accomplished for the first time.

Two methodical historical approaches to creation of artificial laboratory ecosystems with closed or quasi-closed substance circulation have been established. One of them consists of isolation of a biocenosis from its natural habitat, along with a part of its environment. The species composition of such a biocenosis, as a rule, is unknown with the exception of one or more species, — the object of study. Such systems often contain both unicellular and multicellular organisms, including fish, animals and man. Another approach is synthesis of a system from a small number of preliminarily identified plant and animal species, and man, with concomitant microbial cenoses capable of occupying mutually complementary ecological niches and creating closed substance circulation.

Among the ecosystems of the former type, the project "Biosphere-2" occupies a special place. Its framework is composed of a considerable area (about 1 ha) of dry land surface where natural (but unidentified) biocenoses are isolated. The enormous complexity of "Biosphere-2" does not allow us to hope for an exhaustive investigation and understanding of population dynamics and mass interchange of the biocenosis. However, the information about even a part of the species in the total mass interchange of the most important biogens-oxygen, carbon and nitrogen — has an unquestionable scientific value. At the same time, the uniqueness of the "Biosphere-2", and its complexity comparable to that of large natural ecosystems, make each experiment within it also unique and long-lasting. This in turn slackens the pace of accumulation of ecological knowledge.

An example of the other type of ecosystem is the complex "Bios-2" developed and constructed at the Institute of Biophysics from 1969—1971. Its size is by about 700 times smaller than that of "Biosphere-2", and amounts to about 300 m^3, and includes the life unit with cabins for the testers, phytotrons for photosynthetic regeneration of the environment with a sowing area of up to 60 m^2, a 360 kw system for intensive illumination of plants, an interlink gas- and water interchange, water conditioning, information collection systems, etc. The circulation rate in such systems is accelerated many times more quickly than the biosphere.

This allows various possibilities for experimental ecology and physiology, and for simulation of population genetic and evolutionary processes in a small biosphere, since it is transparent and visible in all of its parts, and accessible to analysis and intervention. Special attention should be paid to the role of man in the artificial closed ecosystem. Here, man not only represents a metabolic link, his metabolism involved in the system, but also consciously regulates the pattern and intensity of his relations with the environment, as well as operating the ecosystem. In the Earth's biosphere there are no analogues of this function. The Earth is not controlled by anyone, and its stability is based on stochastic mechanisms, ensured by the enormous size of the planet, and by the diversity of its living organisms. Man's involvement in the closed system makes it possible to detect all his metabolic links with the environment, especially infinitely small links, whose effect is only manifested in circulation cycles.

This possibility must first of all be used for physiological and biochemical studies of the external metabolism of normal and sick people and possibly employed therapeutically, for investigation of adaption to conditions created in the system by the will of the experimenter, and therefore fully controllable. Developing further the topic of the applied use of closed ecosystems including man, one should mention, that the knowledge and the technical solutions obtained in long-term experiments on life support, apart from application in research of space, can be successfully used for a goal-oriented correction of the environment in a real environment.

During the series of experiments from 1972—1984, crews consisting of 2 or 3 testers selected by a medical commission from volunteers, lived in "Bios-3" for 4 to 6 months. The testers independently performed all life support functions: they sowed plants, looked after the crops, harvested and processed the crops, and repaired the technical units and ruled the complex. It is in this respect that "Bios-3" differs from other experimental systems, where the operation is performed from the outside by attending personnel.

The life unit's air which was continuously replenished with carbon dioxide and other volatile products, for the testers' vital activity, was circulated to phytotrons by means of forced ventilation. Here, it was enriched with oxygen from plant photosynthesis, and then was supplied to the life unit again. However, the plants cannot completely purify the air from all its contaminants. That is why a thermocatalytic filter functioned in the gas flow, and burned organic admixtures. The condensate of the moisture released by the plants was the source of water. For drinkable water, the condensate was purified with ion-exchange resins and corrected by means of adding mineral salts. For hygienic purposes it was only sterilized by boiling.

The crew communicated with the outer world by telephone or through "Bios-3" illuminators, which attenuated the psychical strain and diminished the sensation of isolation. Their free time could be spent in individual cabins, in which case the control was performed only through electrophysiological monitors. The life in "Bios-3" over a period of 6 months did not bring about any negative changes in the personnel's health conditions, including the microflora of skin, mucosae and intestines, or any allergic reactions to plants. The atmosphere, water, and vegetable food, repeatedly regenerated in the biological cycle of substance circulation, turned out quite fit for human life support.

The investigations of the Institute of Biophysics pioneered closed ecosystems in 1964. This situation remained unchanged and is recognized by scientists of other countries. From the 60s up to mid-80s, similar research was carried out, at the Institute of Medical and Biological Problems, in Moscow. And since the 80s, experimental approaches to this problem have been intensely studied in other countries (Japan, Canada), but mainly in USA, to create systems for long-term human life support in space (the NASA program) and ecological modelling (the project "Biosphere-2"). In the CES program, experiments with man's inclusion in the system are planned for 1998—1999 and we seek the pursuance of cosmic goals.

The first 2-year experiment in "Biosphere-2", with a crew of 8 testers, and thousands of plant and animal species, was completed in September 1993. In the process of its performance, a unique material was obtained which is now being treated and analyzed. However, one cannot be sure that its analysis will get us closer to understanding the mechanisms of stability and functioning of closed ecosystems, because of the uncontrollable energetics of the system (natural solar illumination) and the enormity of cause-effect interactions which determine the course of ecological events in "Biosphere-2".

At the same time, experiments with the system "Bios-3" permit us to distinguish two kinds of obstacles on the way to ulterior closing of ecosystems: technical and biological obstacles. Although the technical factors do not seem principal, experience has demonstrated that it is only the question of the technical reliability of the system that determines the true degree of its autonomy. In a biological-technical system, where the living and the mechanical parts of the systems are as close as if they were organs of one body, the total life span is determined by the most short-living element. It is, as a rule, the technical part, and it is this factor that will first limit the life time of autonomous life support systems. Here, too, one of the trends of international co-operation can be seen — increased reliability and life time for the physico-technical component of a closed ecosystem.

Among the biological causes hindering a further increase in the degree of closed substance circulation, one should mention the important following events.

The first — coupling of the alimentary input of the human link and kinds of biomass reproduced by the system. First of all, it is not clear what the so-called "optimum ration" must be. It is quite probable that, being composed, taking into account the complete list of substances indispensable for man, it may in the long run turn out to be not optimal, since it may leave in disuse the adaptive mechanisms of the human organism. Besides, there are many reasons to think that the assortment of alimentary plants and animals, (rather narrow compared to the species diversity of the biosphere), has been selected not because of its uniqueness and indispensability, but for economical reasons. Recognition of this opens wide opportunities for search of new food sources among plants and animals which will be more suitable for intensive cultivation, under the conditions of artificial ecosystems, than traditional agriculture items.

The second difficulty of the food problem is the impossibility of selecting one species, which, being a metabolic antipode of man, would reproduce all his food needs, and feed on his excrements. One should cultivate the system's multispecific cenoses and thereby complicate its structure. This results in an increase of the system's size and the consumed power. Besides, a potentially "deadlock" product, in the form of non-edible biomass is formed whose weight is comparable to that of the edible biomass. The problem of its reentry into the circulation so far has no satisfactory solution, and a search for it must still be conducted.

Still another task that has emerged in experiments with the ecosystem, is ensuring its stability. Apart from the technical questions considered above, two biological problems have appeared — microelements and concomitant bacteria.

Migration of microelements in an experimental ecosystem turned out to be a nearly incontrollable process, since they migrate into the substance circulation, from the construction materials of the shell, and are then ingested and concentrated by various inhabitants of the system, to a different degree in different parts of the body, at different periods of life, and under different conditions. It should be underscored that the microelements problem is not technical, although they come from the system construction. The kinetics of their release is determined not by a passive washing-out, but by multiform biochemical activity of the system's microflora, which shows a versatile aggressiveness. Further movement of the elements involved in circulation is also determined by biological processes in the system.

As far as the microbial component of the CES is concerned, there has been failure to stabilize the species' composition and number of microorganizms at any definite level in any of the experiments. The causes thereof are far obscure so far and we are faced with a further search for ways of controlling the system's microflora, since among the present species, there may exist species potentially pathogenic for man.

Now we have to ascertain that experiments on closing the ecosystem have detected a menace to its equilibrium. A search for ways of controlling the kinetics of microelements and the microbial cenoses of a closed ecosystem, as the main factors provoking instability of cycles, is a task for the near future. Taking into account the more than 20 year-long experience of the Institute of Biophysics, the results achieved in the field of creation and investigation of closed ecological systems and the general importance of this problem for mankind, the Presidium of the Siberian Branch of the Russian Academy of Sciences, after Krasnoyarsk was opened, set up on its basis an International Center for Closed Ecological Systems (ICCES). Since it possesses the status of an open laboratory, like an administrative unit, it is possible for foreign institutions and for single scientists to join the ICCES within the framework of joint projects. This seems expedient for optimum acquaintance with, and use of, the accumulated knowledge of the international scientific community.

However, such an interaction generates some problems associated with handing-over and diffusion of information of commercial interest. So far, our foreign partners prefer short-time excursions or inviting single promising scientists. This is no doubt a cheaper way of obtaining information, but it is not efficient because the basis of fundamental and applied research of different countries remains therein unchanged, and no mutual enrichment of social and psychological foundations of the investigation process in different states takes place.

It is obvious that in the framework of international co-operation, the greatest effect could be achieved by means of carrying out a wide-scale long-term experiment, or a series of experiments. With respect to CES, this could be an experiment on closing an international crew in a modernized "Bios-3" system, for a time of up to half a year. Therein, the group of external control and support could also be international and based on a shared stock of analytical equipment. Of course, performance of such experiments requires a certain audacity, and an open mind to overcome the political prejudices and take risks under unstable political and economic condition. It seems that some time must pass before the situation becomes favorable for solution of all the problems and for development of deeper interactions.

Usually, the prospect of an idea concerning life support is determined by its comparative energy and weight characteristics, with the view of satisfying human requirements. However, the life support problem has still another aspect which so far enjoys less attention. This is the problem of characteristics of closed systems, which determine the degree of their isolation from the environment, and therefore the degree of protection of the latter from contaminations with products of vital activities of the systems themselves.

A life support system, based on biological substance circulation, as experiments have shown, is truly feasible and possesses the potential for further improvement, even a capacity for ensuring man's biological expansion beyond the biosphere along with a system which allows his existence, taking from the outside only energy, and controlling it from inside, without releasing any alien metabolites into the environment.

Such an elementary ecosystem, ruled by human intelligence, and provided with knowledge about the relationships between processes in the environment, is just the perfect form for the noosphere to start its expansion into space without threatening other Solar system bodies with irreversible changes, caused by the invasion of terrestrial substances and processes.

This problem of astronautics, the problem of creation of a human life support system, may be defined as that of creation of an autonomous sphere which would optimally satisfy human needs in a maximally economic way, without pollution to the any environment. But is this not the most important task of mankind on Earth? It is in the similarity of these tasks that we see the greatest possibility for creation and improvement of artificial ecological systems for man, and the greatest incentive for joint efforts of scientists from different countries and peoples.

♦

GENETIC EFFECTS OF RADIATION AND OTHER MAN-MADE POLLUTION IN SIBERIA

V.K. SHUMNY, N.A. KOLCHANOV, N.N. DYGALO, V.G. MATVEEVA, L.N. IVANOVA, M.I. VOEVODA, A.G. ROMASHCHENKO, E.I. KARAKIN, I.I. KIKNADZE, E.B. BUDASHKINA and S.E. PELTEK
Altai International Centre for Humanitarian and Biospheric Research, Institute of Cytology and Genetics of SB RAS, Novosibirsk, Russia

ABSTRACT. Genetic consequences of nuclear tests in the Semipalatinsk region and other man-made pollutions on populations and ecological systems in the Altai territory were studied. The genetic and functional abnormalities of human, animal and plant communities of the Altai territory were found. Test systems and methods used were proved to be sensitive for assessment of effects of man-made pollutions. Further studies under natural conditions and model experiments are necessary for elucidaing of the nature of detected genetic deviations and establishing of their relation to the radiation agent.

Keywords: nuclear tests / man-made pollution / genetic consequences / human populations / animal and plant communities.

A challenging problem of world ecology lies in elucidating the consequences of man-made pollution of the environment for the population of our planet. Pollution resulting from nuclear weapon tests are hazardous to the present and future generations.The aim of the first step of the study was the development of test systems and methods for assessment of the genetic consequences of nuclear tests in the Semipalatinsk region and, in broader terms, assessment of the genetic consequences of the effects of man-made factors on populations and ecological systems in the Altai territory. This problem is solved in several exemplary districts of the Altai territory, in which there are two aspects:

- cytogenetic, genetical-physiological and molecular-genetical studies of populations from different districts of the Altai territory;
- assessment of the genetic load posed on natural populations of plants and animals.

Human populations

The study was performed with the aim of the identification and analysis of permanent residents of the Altai territory exposed to nuclear tests and radiation fallouts during nuclear weapon tests in 1949-1963. In the Loktevsky and Talmensky districts, total morbidity of children at the age

V A Koptyug and J Klerkx (eds), Science Policy New Mechanisms for Scientific Collaboration between East and West, 69–80

of one year was found to have increased 7-fold in 1989 (A.S. Obert *et al.*, 1993). Predominant among the disease patterns was congugation gaundice (hyperbilirubinemia, HBN) developing during the first days after birth and not conforming with the conventional "neonate gaundice" (Lupandin, 1993) in different associations affecting the central nervous system. Among adult populations of the radiation-exposed territories oncological diseases, particularly "atypical" variants, for example, cancer of lungs and thyroid gland in men from rural communities, predominated. The frequencies of leucoses and anemias refractory to treatment increased significantly (V.G. Matveeva *et al.*, 1993).

The residents of the town Gornyak of the Loktevsky district located 150 km from the nuclear testing grounds were examined (total number 136 individuals). Newcomers who were reliably exposed to different doses of radiation during tests on nuclear grounds (5 men), and people who had suffered from the Chernobyl catastrophe served as positive controls. Known data in the literature were used as negative control. The material for chromosomal analysis was as follows. PHA-stimulated mitotic cells from peripheral blood or bone marrow cells. Lymphocytes were cultured by the standard micro and macro methods. Chromosome preparations were made and analysed conforming to international standards (Ford, 1983). Two hundred metaphase plates, on the average, were analyzed for each individual. The entire spectrum of aberrations of chromatide and chromosomal types was taken into account (Radiation-induced..., 1983). The so-called "gaps" and aneuploid cells were discounted. Chromosome damage and karyo-type abnormalities were identified with the use of G-banding.

14 children ages 6 days—8 years, had gaundice (HBN), and in their mothers ages 19—36 years, the total percentage of cells with aberrations was 1—6%, 3.1% on the average, which exceeds twofold the maximum value accepted as international control (D. Anderson *et al.*, 1988). Aberrations of chromosomal type predominated in the majority of children and particularly in their mothers (D.L. Preston *et al.*, 1989; F. Vogel, 1992). There were 1 dicentric + ring (d+r) per 2.000 metaphases in the negative control. In contrast, we identified 1 d+r per 400 metaphases in this sample after 48 hours of culturing; in 9 children ages 7 days — 11 years with birth defects, and in their mothers ages 19—41 years the total percentages of aberrations ranging from 0.5—7.0%. The average percentage for children was 2.9% and that for their mothers was 3.6%. Aberrations of chromosomal type predominated. Single cells with conspicuous multiple breaks and exchanges occasionally occurred. Furthermore, we detected a child with Down's syndrome and also two children with balanced translocation of familial+ type 46 XX, t(17,22) and 46 XY, t(15p$^+$).

We examined 42 individuals from 3 generations of 8 families in which there were individuals with high frequency of chromosomal aberrations. A 1.5 year old boy, a carrier of an abnormal clone was identified. The time course of this clone was dramatic, initially it composed 12% of total cell number and 26% of this number after 4 months. This boy soon died from acute lympholeucosis. 3.3% of cells from his 23 year old mother showed chromosomal aberrations, half of which were of chromosomal type; from his father's pedigree it is seen that his 53 year old grandmother had the highest percentage of chromosomal aberrations, and from the mother's side, his 46 year old grandfather had 4.4% of these aberrations. This pattern was the one typically observed in the majority of families.

Seventeen miners were included in investigation. Their ages ranged from 30—50 years. They worked in mines for 10 years, on the average. They have been in contact with heavy metal salts during all these years. This sample was taken to differentiate the damaging cytological consequences of chemical and radiation mutagens. Single dicentric and ring chromosomes were identified only in 5 miners, fragments were identified in the remaining miners, and this is twice the number for the controls. The low percentage determined for miners may be the result of the weak effects of chemical mutagens or viruses (Froland, 1962; ISCN, 1978).

The total number of aberrant cells, those of cells with chromosomal, and those of cells with d+ ring type for the groups compared in the histogram (Fig. 1). In spite of the differences in percentage, there are obvious similarities in cytogenetic effects.

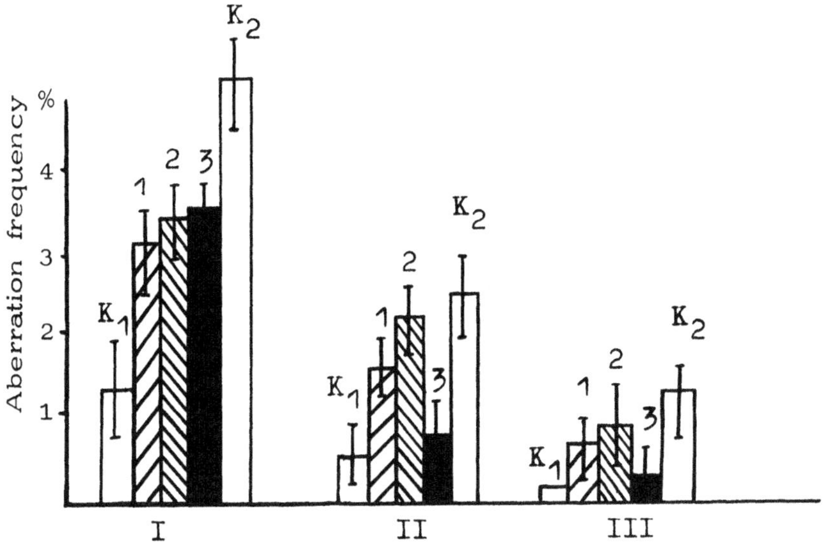

Fig. 1.
Mean frequencies of chromosome aberrations in residents of town Gornyak, Altai territory.
I — total aberration frequency;
II — aberrations of chromosomal type;
III — frequency of dicentric and ring chromosomes;
K_1 — negative control; K_2 — positive control;
1 — children with HBN and BD; 2 — their mothers; 3 — miners.

In "yellow" babies with specific gaundice (hyperbilirubinemia, HBN), anemia, leucopenia, retarded development, a set of neurological disorders persist for several years (L.N. Ivanova *et al.*, 1993). The pathogenesis of this syndrome is not clear. However, when taking into account the variety of abnormalities and the manifestations in the second generation, it appears that neuroendocrine mechanisms are involved in developmental deviations under the effect of deleterious agents acting during early critical periods. Significant here may be changes in the activity of the thyroid gland involved in the regulation of morphogenesis and also in the adrenal corticoid function providing adaptive responses.

We studied 88 families in the Loktevsky district having at least one child with prolonged gaundice born in 1989—1992, and 15 families in which all the children were considered to be healthy. Blood and urine samples were collected from each subgect (mothers, fathers and their children) of the studied family. The standard radio immunoassay was used to measure the concentration of triiodthyronine (T_3), thyroxine (T_4), thyroxinebinding globulin (TBG), and plasma cortisol. Determination of the free cortisol in urine was performed by the fluoremetric method (Bunyatyan, Ignatenko, 1985). Student's t-test was used to test the significance of the data.

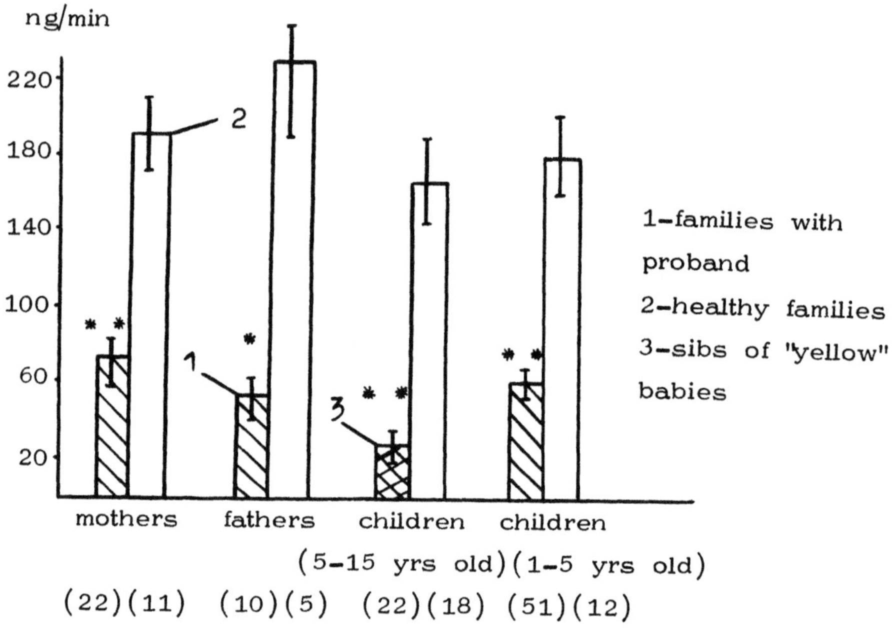

Fig. 2.
Excretion rate of free cortisol in urine adjusted to standard body surface (1.73 m^2) in parents and children in families with and without "yellow" babies.
() — number of subjects studied. Student's "t"-test, * — $p < 0.05$; ** — $p < 0.01$.

It was shown that decrease in the function of the thyroid gland is characteristic of mothers giving birth to "yellow" babies. Levels lower than normal of T_4 and TBG were determined in fathers. In contrast, T_3, T_4 and TBG concentrations in "yellow" babies were found to be higher than normal. The values for thyroid function were in the normal range or close to it in mothers and fathers of healthy families living in the same district. In healthy children, the concentration of T_4, the major hormone reflecting the level of thyroid secretion, was significantly higher than in "yellow" probands, i.e., falling within the normal range. The excretion rate of cortisol with urea reflecting the secretory activity of the adrenals was found to be much lower in families with "yellow" babies (Fig. 2). Of particular interest are comparisons of the endocrine status in sibs of "yellow" babies without HBN in early childhood with their age counterparts from healthy families. It was found that excretion value of cortisol was significantly lower in "yellow" babies than in their sibs, and in sibs of "yellow" babies significantly lower than in counterpart children in healthy families.

Thus, significant changes in the function of the thyroid gland and adrenals were detected in parents and children regardless of gaundice manifestation in families from Loktevsky district with "yellow" babies registered in pedigrees. Taken together, our observations allow us to suggest that hyperbilirubinemia in the polluted Altai may be one of the pleiotropic effects of somatic mutations manifesting as a impairment of systems controlling neuroendocrine mechanisms of ontogenesis.

Use of PCR-based techniques for sequencing of DNA has enabled us to quantitatively estimate the effects of radiation and other man-made polluting agents on the genetic apparatus (M. Litt, J.A. Luty, 1989; M. Orida *et al.*, 1989; T.J. White *et al.*, 1989; M. Hoss *et al.*, 1992). To solve a problem, which includes evaluation of the deleterious effects on the degree of damage of the genetic apparatus in residents of the Altai territory, the following experimental scheme is suggested. The choice of genetic loci to be analyzed is based on data on the high sensitivity of the hemopoietic system to radiation. These are the genes of two receptors for the macrophage colony stimulating factor (M-CSF) and the granulocyte macrophage colony stimulating factor (GM-CSF). Changes in the structure of the receptor genes and in their expression level can produce significant physiological deviations (C.S. Sherr *et al.*, 1985; K.G. Peters *et al.*, 1992) from the norm. The M-CSF and GM-CSF receptor genes are good targets owing to their great number of loci with mutational "hot spots". We also analyzed the hypervariable region of mitochondrial DNA, the most thoroughly studied DNA for many human population in the world (M.I. Voevoda *et al.*, 1993).

We chose for analyses the German-origin population from two districts of the Altai territory, control Nemetsky (German) district and radiation-exposed (Rubtsovsky). The Rubtsovsky district was represented by 41 families including both parents and children, in addition, 27 pairs of relatives providing information about hereditary transmission of changes in mtDNA and nuclear genes. Forty eight families and 19 pairs of relatives were examined in the Nemetsky district. Taking into consideration familial relations of the second degree, the studied samples allow to form familial structures of the three generations and also to identify additional informative pairs. Based on the criterion of residence in the Rubtsovsky district, all the examined individuals were regarded as exposed to the action of radiation for different periods of life in the Nemetsky district, radiation exposure was excluded.

Blood analysis in the examined population included counts of the number of erythrocytes, leucocytes, trombocytes, determination of hemoglobin values and isolation of DNA from venous blood. As a result, a bank of more than 500 samples of DNA from residents of the Altai territory was established. The data of clinical investigations, genealogical information, blood parameters, residence duration and all the relevant data were computerized.

From the results of analysis of parameters of peripheral blood, it follows that Rubtsovsky residents have deviations in all types of myeloid cell lineages compared to Nemetsky residents. Abnormalities were particularly obvious for erythropoietic cells. The degree of the observed deviations correlated, as a rule, with duration of residence in the exposed territories. The mean erythrocyte number was significantly lower in the Rubtsovsky district for both sex groups. The same trend was observed for hemoglobin level in Rubtsovsky residents (Fig. 3). For individual Rubtsovsky residents, relatively high lymphocyte level and lower relative and absolute levels of cells of neutrophils and eosinophils were observed. There was a significant decrease in leukocyte number with increasing total residence in the given district as well as duration of residence from 1949—1962.

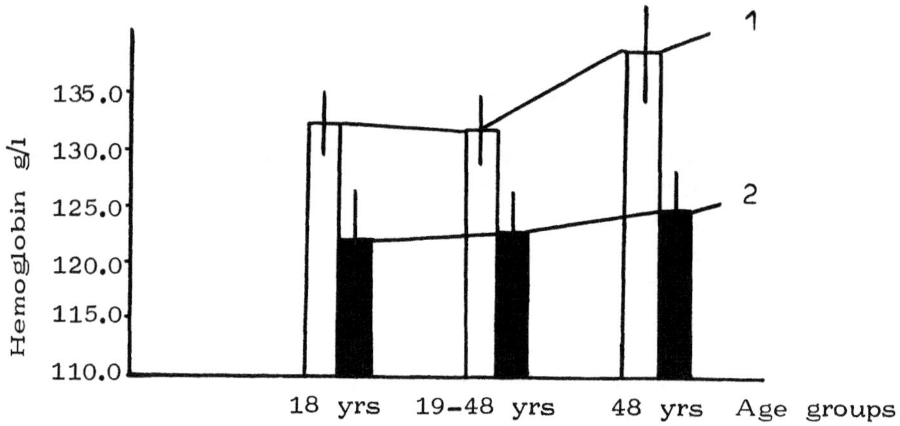

Fig. 3.
Age changes in hemoglobin values in women.
1 — Nemetsky district; 2 — Rubtsovsky district.

The pattern of changes in parameters of the peripheral blood in residents of the Rubtsovsky district generally resembles the symptoms of radiation damage involving the erythrocyte link and related to residence duration in the exposed district. Based on peripheral blood parameters of Rubtsovsky residents a DNA bank was established which is currently unique material for detailed molecular-genetic studies. The size of family samples included in this analysis is large enough to estimate spontaneous mutation at chosen loci. This allows us to rely upon the possibility of evaluating the radiation induced changeability according to genetic loci.

Rodents

Wild *Rodentia* populations were studied in three radiation-exposed districts of Altai (Rubtsovsky, Loktevsky, Uglovsky), and a fourth district (Krasnogorsky) not exposed to radiation served as control. The blood composition in mice from control district does not differ from the one observed in laboratory mouse stocks. Abnormalities in blood composition were the highest in mice from the Loktevsky district. Analysis of patterns and levels of chromosomal aberrations in bone marrow and spleen cells in the *Rodentia* species studied demonstrated predominance of chromatid and chromosome bridges (60—98%), single or multiple chromosome fragments occurred less frequently. The lower-ranges of chromosomal aberrations are almost the same in all the four Altai districts, (2.0—4.8%), and their upper ranges in

Rubtsovsky and Loktevsky districts were within 18.0—23.4%, approximately twofold higher the upper range in the Krasnogorsky district, 10.8%. Of 206 animals exposed to total pathomorphological screening, neoplasms were detected in 13 animals regardless of sex and species. In the samples analyzed, cancer frequencies were 10.2% in Loktevsky district and only 2.8% in control Krasnogorsky district. The discovered trends suggest that there may be a genetic component in predisposition to blood abnormalities, chromosomal aberrations and cancer diseases of a specific type in the district (E.I. Karakin *et al.*, 1993).

To elucidate the possible nature of genetic and physiological abnormalities in populations and residents of the Altai, an experimental analysis of the effect of known radiation doses covering the range calculated for the exposed territories was carried out. External and internal exposure of female rats during the second half of pregnancy to gamma rays and to a radionuclide spectrum similar to the fallout over Altai territory (total doses were 0; 15; 105 or 300 rad; 117 female rats and their 760 offspring) reproduced some cytogenetic effects and affections of the physiological systems and among them hematology, endocrinology, reproduction, child-bearing mentioned above and observed by clinicians in populations of Altai in the range of the Semipalatinsk test grounds. A bank of DNA samples from generations of these rats is a material for detailed molecular-genetic studies of the effects of radiation on the genetic apparatus. The future prospects is monitoring of cytogenetic, biochemical, physiological and molecular-genetic effects of radiation through generations.

Chironomids

The larvae of chironomids live and burrow in sediments concentrating radionuclides and heavy metals and, hence, they are subgect to pollution effects. The giant polytene chromosomes of chironomids allow to perform high resolution cytogenetic analysis and to estimate radiation effects. Analysis of karyotypes and chromosomal polymorphism was carried out on squash preparations of salivary glands of chironomids of the Chironomidae subfamily collected in 1991—1993. The collection sites were known to be radiation-polluted: Chelyabinsk region, the territories polluted as a result of the Kyshtym catastrophe, the Altai territory polluted as a result of Semipalatinsk nuclear tests. The unpolluted "clean" Altai districts, European part of Russia, Far East, West and East Siberia served as controls (I.I. Kiknadze *et al.*, 1993).

Comparative analysis of banding patterns in *Camptochironomus tentans* demonstrated increased level of chromosomal polymorphism in populations from polluted Altai and Ural regions. Evidence for this were:

a) seven new unique inversion sequences of the bands tenB7, tenB8, tenB9, tenC4, tenD5, tenD6, tenE8 detected in chironomids from the Chelyabinsk reservoirs and those from lake Noven'koe in Altai;
b) increase in number of genotype combinations of inversion sequences;
c) increase in the average number of heterozygous inversions per individual compared with the other populations.

Factors producing chromosome variability may be radionuclides. An increased level of chromosomal variability is characteristic of *C. tentans* larvae from the lakes of the Chelyabinsk region ("Mayak" territories) where the radioactivity level is significantly higher than the natural background level. The same was observed in the Noven'koe lake (Altai territory) in which the present radionuclide content does not exceed the one accepted as natural background. However, radionuclide concentration twice exceeds the natural background level in sediments at depths of 4—10 cm, this may evidence for the sediments being radionuclide-polluted many years ago. If this was the cause of the increased level of chromosomal variability in this particularly reservoir, the above becomes explicable.

Plants

Mutational events induced in plant communities by nuclear tests can be identified with the use of modern cytogenetic and biochemical methods. In collection of samples of wild plants, we took into account the information derived from medical-demographic accounts summarizing previous studies in Altai territory; based on this information, the Uglovsky, Loktevsky, Rubtsovsky districts were considered as the most endangered in terms of population morbidity (V.K. Shumny *et al.*, 1993).

The first mitosis in meristematic plant roots of perennial plant species and microsporogenesis in pine and *Hordeum bogdanii* were examined. They were under examination during meiosis, the key step of plant development. Cytogenetic analysis was performed by the standard methods (Pausheva, 1970). The abnormalities taken into account included bridges, fragments, chromosome lagging, micronuclei, asynchromatic division. Differences in frequencies of abnormal cell division were found between wild populations of *Agropyron cristatum* in unpolluted (0.16—0.35%) and polluted (1.2—7.2%) districts during the first mitosis. The main abnormalities were lagging chromosomes and fragments.

We chose pine in our studies because this particular woody plant has been growing since the first radiation exposure in the Altai territory. In pine, the identified abnormalities of meiotic division resulting in complete or partial sterility may be evidence for long-term effects of radionuclides. Studies of meiosis in pine showed that the plant populations in polluted districts considerably exceeded the control group in the frequency of anomalies of cellular division. The percentage of cells with abnormalities was 35—40% in some pines. The main abnormalities were asynchronous disjunction of chromosomes and also abnormal distribution of chromosomal material. The abnormalities resulted from impairment of mechanisms controlling spindle division and they result in aneuploidy of gametes, and, consequently, sterility. It should be emphasized that no trees with meiotic abnormalities were identified in the Kulundinskoye forestry (Novosibirsk region) taken as control. Thus, comparisons of wild grass and woody species collected from polluted and unpolluted districts of Altai territory revealed differences in the frequencies of abnormal cell division at mitosis and meiosis.

To analyze storage proteins, we used our previously developed method of separation in 10% acrylamide gel in the presence of high urea concentration (S.E. Peltek *et al.*, 1984). Program packages Quattro were used for computer treatment of the data. The electrophoretic pattern of grass crested wheat (*Agropyron cristatum B.*) seed storage proteins consists of 10—20 components. The total number of components in all the populations was 46. The electrophoretic mobility of each component was compared with standard (wheat gliadin), and the comparative results are now stored in databank. The frequency of 9 of the 46 components, which were identified in all the populations, is almost 0 in the control population (Cherga). Components with frequencies close to 0 were much less frequent in the other populations. The data indicate that the total spectrum of frequencies within each *A. cristatum B.* population is formed independently, and polymorphism for Gornyak, Topolnoye and Rubtsovsk populations is significantly different from the one for Cherga population. It should be noted that *A. cristatum B.* is a perennial cereal, and, for this reason, the contribution of mutant variants of storage proteins, which is not immediately discriminated by natural selection, increases with successive generations.

We started this analysis of electrophoretic patterns of Scotch pine (*Pinus silvestris*) storage proteins, being aware that this woody plant is a "witness" of an ecological catastrophe. In contrast to *A. cristatum B.* in which mutations have been under natural selection for generations, pine enables us to analyze mutations in somatic tissues. The electrophoretic patterns of pine storage proteins consist of 5 well separable fractions, of which 4 are monomorphic and 1 (the fourth) is polymorphic with a fast and a slow variants. Along with polymorphism for the fourth fraction, additional components and null variants of all fractions occur in pine populations. The fast variant of the fourth fraction is more frequent in Volkovo, Maslanino, Pogranichnoye and Laptevskoye populations, and the ratio of the fast to slow variant is approximately 1 for Topolnoye and Cherepanovo populations.

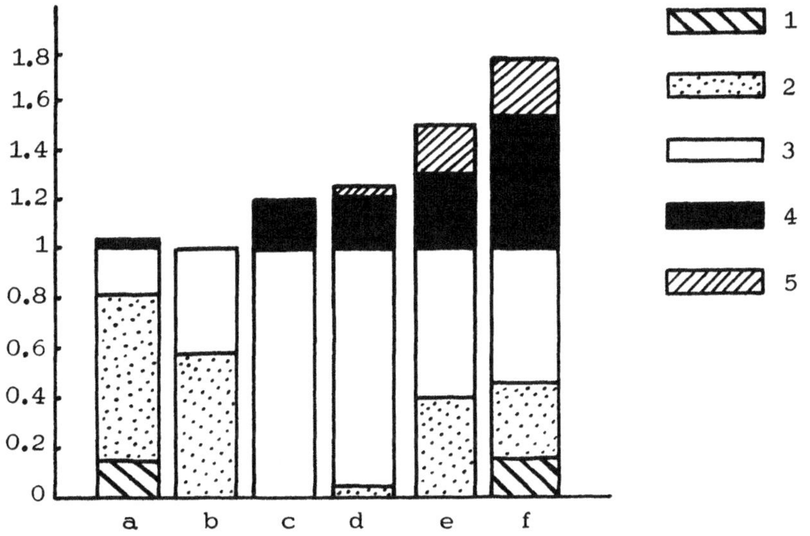

Fig. 4.
The frequencies of individual components of seed storage proteins in Pinus silvestris.
Control: a) Volkovo; b) Maslanino; c) Cherepanovo;
Analysed: d) Topolnoye; e) Laptevskoye; f) Pogranichnoye.
1 — slow variant of fraction 4; 2 — fast variant of fraction 4;
3 — heterozygotes(*) by fraction 4; 4 — additional components; 5 — null variants.

Fig. 4 presents the results of analysis for storage proteins of pine seeds in populations from Altai territory with the Novosibirsk region serving as control. A column of the histogram rep-

* — "Heterozygote" is a conventional symbol for individuals having two variants of fraction 4.

resents a population, the height of the column is the sum of the frequencies of components additional to the major variants of storage protein pattern, null variants, slow and fast variants of fraction 4 and frequency of occurrence frequency of both variants in an individual. A part of the column above 1 is the sum of null and additional variants frequencies. As shown in Fig. 4, mean heterozygosity of a population is not directly related to the ecological conditions of the population. In fact, the content of additional components and null variants in Pogranichnoye and Laptevskoye populations is significantly higher than in the other populations, and, thus, it may be evidence for higher concentrations of mutations in these populations.

The data testify that the genetic structure of ecologically stressed *A. cristatum B.* populations is completely different from the one observed under natural conditions. The differences concern mainly larger variations in seed storage proteins. The results of analysis of pine seed storage proteins demonstrate the accumulation of mutations produced by both the short and long-term effects of man-made pollution.

The results of the studies presented in this paper provide evidence indicating that there are genetic and functional abnormalities in a set of examined characteristics of human, animal and plant communities of the Altai territory. It is quite obvious that elucidation of the nature of the detected genetic deviations and establishment of their relation to the radiation agent is a problem by far more complex than detection. Further studies under natural conditions and model situations are needed.

References

1. Anderson D., Jenkinson P.C., Dewdney R.S., Francis A.J. (1988). Chromosome aberrations, mitogen-induced blastogenesis and proliferative rate index in peripheral lymphocytes from 106 control individuals of the U.K. population Mut. Res. 204:407—420.
2. Bunyatyan A.F., Ignatenko L.L. (1985). Extractionless method for determination of free corticosteroids in the urine based on competing binding of transcortine. Lab. Work. 1:24—27 (in Russian).
3. Ford E.H.R. (1983). Human chromosomes. Acad. Press. London, New York.
4. Froland A. (1962). A micromethod for chromosome analysis on peripheral blood-cultures. Lancet. 2:1281—1282.
5. Hoss M., Kohn M., Paabo S., Knauer F., Schroder W. (1992). Excrement analysis by PCR. Nature. 359:199.
6. ISCN (1978). An international system for human cytogenetic nomenclature. Cytogenet. Cell Genet. 21:309—404.
7. Ivanova L.N., Benimetski Yu.S., Gritsko N.G., Melidi N.N., Shulga V.A. (1993). Assessment of the functional status of thyroid gland and adrenal cortex in children with hyperbilirubinemia from the "nuclear trace" region (Altai, Loktevsky district). In: "Genetic effects of anthropogenic environmental factors". Issue 1. pp. 34—44. Novosibirsk.
8. Karakin E.I., Semenova L.A., Palamodova L.A., Paymuratova T.V., Slynko N.M., Gulevich V.V., Leonova I.N., Osadchuk A.V., Lutzenko N.D., Kutuzova T.V., Ronichevskaya G.M., Zvereva L.N., Nikolin V.P., Popova N.A., Ilnitzkaya S.I.,

Serova I.A., Cheresiz S.V., Bochkarev M.N. Genetic monitoring-92 of wild populations of Rodentia from the Altai territory. In: "Genetic effects of anthropogenic environmental factors". Issue 1. pp. 81—91. Novosibirsk.

9. Kiknadze I.I., Istomina A.G., Gunderina L.I., Aimanova K.G., Filippova M.A., Siirin M.T., Sobanov Yu.V. (1993). Cytogenetic monitoring of natural chironomid populations in Altai under conditions of man-made pollution. *Ibid.* pp. 63—80.
10. Litt M., Luty J.A. (1989). A hypervariable microsatellite revealed by *in vitro* amplification of a dinucleotide repeat within the cardiac muscle actin gene. Ann. J. Hum. Genet. 44:397—401.
11. Lupandin W.M. (1993). "Yellow" children. Nature. 4:11—14 (in Russian).
12. Matveeva V.G., Sablina O.V., Eremina V.R., Biltueva L.S., Rogatcheva M.B., Grafodatsky A.S. (1993). Cytogenetics of birth defects in residents exposed to the radiation in the Altai territory. In: "Genetic effects of anthropogenic environmental factors. Issue I. pp. 5—17. Novosibirsk.
13. Obert A.S., Myachkina T.M., Yakov L.E., Voronova N.A., Filiptsova M.Yu., Solovjov N.F., Radchuk M.G. (1993). Pathology of neonates in the Loktevsky district. In: "Nuclear tests, environment and health of the Altai territory population: Materials of scientific research. v. 5. Book 2. pp. 12—41. Barnaul (in Russian).
14. Orida M., Suzuki Y., Sekiya T., Hayashi K. (1989). Rapid and sensitive detection of point mutations and DNA polymorphism using the polymerase chain reaction. Genomics. 5:874—879.
15. Pausheva Z.V. (1970). A practical handbook for plant cytology. Kolos, Moskva (in Russian).
16. Peltek S.E., Pshenichnikova T.A., Sarapultsev B.I., Maystrenko O.I. (1984). A highly efficient method for separation of wheat glyadin. In: Materials KOS SEV. Issue 9. pp. 121—124. Odessa (in Russian).
17. Peters K.G., Marie J., Wilson E., Ives H.E., Escobedo J., Del Rosario M., Mirda D., Williams L.T. (1992). Point mutation of an FGF receptor abolishes phosphatidylinositol turnover and Ca flux but not mitogenesis. Nature. 358:678—681.
18. Preston D.L., McConney M.E., Awa A.A. (1989). Comparison of the dose-response relationships for chromosome aberration frequencies between the T65D and DS86 dosimetries RFRF TR 7—88. Hirochima.
19. Radiation-induced chromosome damage in man. (1983). Ishihara T., Sasaki M.S., eds, Alan R. Liss Inc., New York.
20. Sherr C.S., Rettenmier C.W., Sacca R., Roussel M.F., Look A.T., Stanley E.R. (1985). The c-fms proto-oncogene product is related to the receptor for the mononuclear phagocyte growth factor, CSF-1. Cell. 41:665—676.
21. Shumny V.K., Budashkina E.B., Zhelesnov A.V., Sidorova K.K., Peltek S.E., Goryachkovskaya T.N., Numerova O.M., Kalinina N.P., Zhelesnova N.B., Shchapova A.I., Uzhintseva L.P., Zagorskaya A.A., Kovalenko V.I. (1993). Assessment of the consequences of the effects of nuclear tests and other anthropogenic pollutions on plants. In: "Genetic effects of anthropogenic environmental factors". Issue I. pp. 91—103. Novosibirsk.
22. Voevoda M.I., Avksentyuk A.V., Ivanova M.V., Lisichenko O.V., Juravskaya E.I., Podkolodny N.L., Kondrahin Yu.I., Naumochkin A.N., Romashchenko A.G.,

Kolchanov N.A. (1993) Population study of the polymorphism of the M- and GM-colony-stimulating factor receptor genes and of mitochondrial DNAs in areas of potential radiation pollution. Communication I. *Ibid.* pp. 18—33.

23. Vogel F. (1992). Risk calculations for hereditary effects of ionizing radiation in humans. Hum. Genet. 89:127—146.
24. White T.J., Arnheim N., Erlich H.A. (1989). The polymerase chain reaction. Trends in Genet. 5:185—189.

♦

INTEGRATED SUSTAINABLE DEVELOPMENT METHODOLOGY

St. BONFA
Environment Research Association for the Mediterranean
Rome, Italy

\BSTRACT. A methodology is proposed for integrated sustainable development based on environmental planning. The methodology is described from the phase of the determination of the objectives to the phase of management of a project.

Keywords: sustainable development / environment / planning / management.

Methodological Approach

The methodology is based on "Integrated Approach", in which environmental planning is seen in terms of a cross-functional process. The regional development plan represents the point of intersection of two planning functions, the vertical and the horizontal (fig. 1).
The vertical function represents the overall environmental planning from the national or macro level to the local or micro level, and coordination between them at an intermediate (regional) level.
The horizontal function covers the whole range of environmental planning activities at the intermediate (regional) level chosen.
The point at which the two funtions cross each other expresses a working synthesis which is the Integrated Environmental Master Planning.
The activities flow chart (fig. 2) indicates the planning activities sequence.

The sequence of activities is as following :

1) Determination of environmental objectives

At this preliminary stage the planners examine the development objectives formulated at the national level, and accordingly define the overall environmental objectives to be obtained within a certain period at both regional and local level.

V A Koptyug and J Klerkx (eds), Science Policy New Mechanisms for Scientific Collaboration between East and West, 81–86

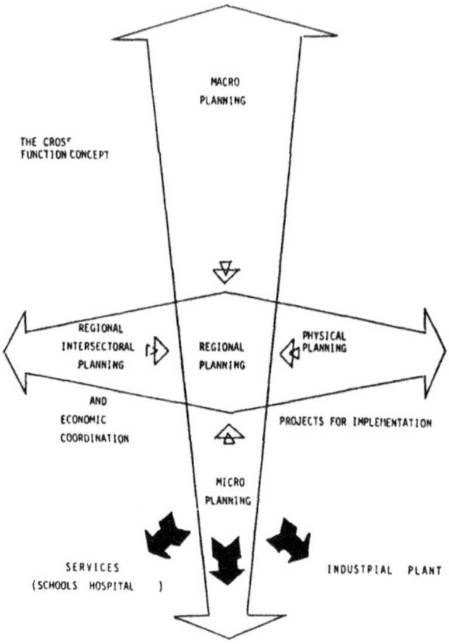

Fig. 1.
Regional development plan

2) Compilation of environmental data and survey of the region

Environmental planning is based as far as possible on reliable actual data obtained from available published material. A careful survey of the region yields additional data in the form of verbal communications and personal observations. On the basis of the accumulated environmental data and detailed environmental maps, the region is divided into homogeneous ecological zones (environmental zone).

The methodology to reach the Environmental Zone is as such :

- (i) formulation and selection of agriculture, industry and services land utilization types (LUTS);
- (ii) determination of environmental requirements of land utilization types;
- (iii) compilation of climatic resources inventory;
- (iv) compilation of soil and landform resources inventory;
- (v) compilation of land use inventory (including socio-economic aspect);
- (vi) compilation of a computerized environmental resources data base in a Geographic Information System (GIS);
- (vii) definition of ecological (environmental) zone using a Geographic Information System and based on land, water and socio-economic evaluation.

This will facilitate to carry out multiple goal optimization to maximize achievement of desired objectives and the subsequent planning of agriculture, industry and services land use and the economic activities sustainable with the environment.

3) Diagnose of existing economic, social and institutional situation

On the basis of the acquired data a description and analysis of the existing economic, social and institutional situation can be performed, such as a diagnosis which will serve as a starting point for the environmental macro plan and later for the various sectorial environmental plans as well.

4) Economic environmental macroplanning

A regional growth model is formulated and the environmental planning experts work out a preliminary plan for distributing the available resources of labour and capital among the different economic sectors and environmental zones.

5) Environmental sectorial planning

An environmental development plan is worked out for each sector. The various sectorial plans are then expressed in physical spatial terms.
A organizational hierarchy can then be worked out for the services.

6) Intersectoral environmental integration of plans (Environment Master Plan)

The environmental sectorial plans are now compared with the preliminary environmental plan for each sector at the macro level.
By making iterative adjustments the planners arrive at a final version of the intersectorial environmental development plan.

7) Definition of environment investment objectives (Investment Program)

The objectives generate and detail the short, medium and long terms (six years) investment program fitted to the objectives stipulated on the Integrated Environmental Master Plan.
The environmental investment program is integrated with the environmental master plan that generates and coordinates comprehensive and environmentally sustainable investment projects.
We are now in a position to propose an organizational framework within which Integrated Environmental Investment Plan Projects are identified and move then to the management and implementation phase.

8) Detailed Environmentally Investment Project Management

Understanding and link the projects planned to the overall development is vital for a successful and environmentally sustainable successful project and subsequent implementation of any technical project or institutional reform.

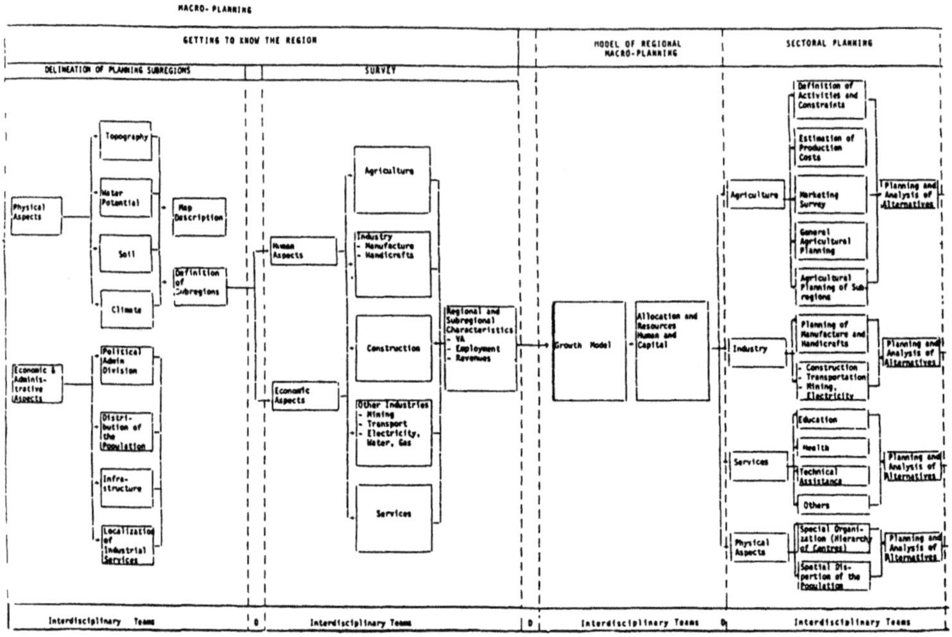

Fig. 2.
Activities flow chart

MICRO PLANNING (IN A SELECTED AREA)

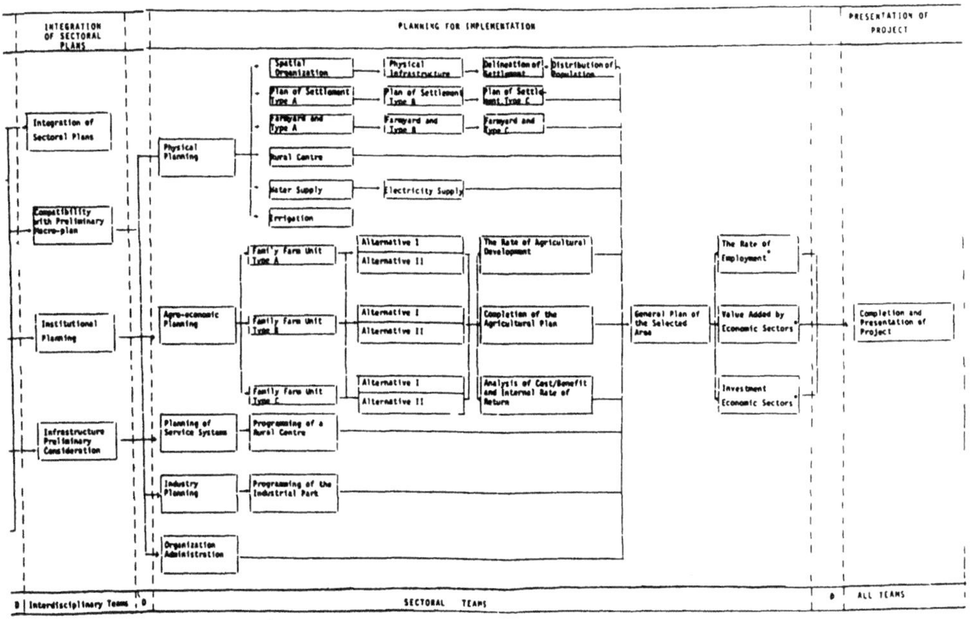

[*] By 5 year stages of implementation
[D] Discussion

A package of integrated investment management projects follows the standard approach:

A) Investment project environmental planning
B) Environmental investment project appraisal and preparation
C) Environmental investment project monitoring and evaluation
D) Environmental investment project implementation and management
E) Investment project financial management.

A - ENVIRONMENTAL INVESTMENT PROJECT PLANNING

Objective : to formulate logically consistent technical investment environmental projects which:

- contributes to identify a development plan objective (step 6),
- have a specific immediate objective which can be shown to contribute to the development plan objective,
- produce outputs which are likely to achieve the desidered immediate objective,
- set out all the activities required to produce the project outputs,
- set out all the inputs which the activities will convert into project outputs,
- set out all the prerequisites and prior-obligations for the success of the project.

B - ENVIRONMENTAL PROJECT APPRAISAL

Objective : to fix the main critical environmental points such as development objectives, immediate objective etc. The appraisal of the environmental project planning should be carried out.

C - ENVIRONMENTAL PROJECT MONITORING AND EVALUATION

Objective : to identify different procedures and techniques of monitoring and evaluation of environmental activities.

D - INVESTMENT PROJECT IMPLEMENTATION

Objective : to organize and to manage investment projects such as : project organization, work breakdown, project control, time management, building teams, etc.

E - INVESTMENT PROJECT FINANCIAL MANAGEMENT

Objectives : to manage or to supervise all investment project operations which involve the use of money.

II.2. INTERDISCIPLINARY STUDIES OF LAKE BAIKAL

BAIKAL INTERNATIONAL CENTRE FOR ECOLOGICAL RESEARCH: PRINCIPLES OF ORGANIZATION AND FIRST SCIENTIFIC RESULTS

M.A. GRACHEV
Baikal International Centre for Ecological Research,
Limnological Institute of SB RAS, Irkutsk, Russia

ABSTRACT. Baikal International Centre for Ecological Research (BICER) founded in 1992 by Siberian Branch of Russian Academy of Sciences, Royal Belgian Institute for Natural Sciences, Royal Society of London, Japanese Association for Baikal International Research Programs, University of South Carolina, has given a significant contribution to successful development of cooperation between scientists of Russia, and of NATO nations. In 1993, foreign scientists spent 2700 man-days in international expeditions on Lake Baikal. The results obtained have been published in best international scientific journals.

Keywords: Baikal / international interdisciplinary cooperation / new science policy / open laboratory.

A few years ago Russia and other countries of the former Soviet Union started an attempt of integration with the community of nations which claim that human rights and economic freedom are their top priorities. This process has brought some benefits to Russian science due to establishment of many new links with the international scientific community. On the other hand, it has up to this date put Russian science on a brink of survival, and some advisers already suggest reducing it twice because, in their opinion, our nation cannot afford to maintain a scientific community of the present size [1]. They say that expenditures for scientific research in Russia should be equal to a smaller fraction of our national income to make this activity more cost-efficient. However, I very much doubt that these experts start from a correct premise: any honest and informed observer will agree that Russian science, taking into account existing salaries and expenditures for purchase of equipment, is by far more cost-efficient than science in any of the NATO countries. For many years young people believed that scientific research is the most honourable field of human activity, and, thanks to some serious, honourable competition for higher education and for jobs in the Academy of Sciences, we now have a valuable national asset - an army of able and well-trained workers in basic research. Another asset is a network of scientific institutes, laboratories, field stations, observatories, scattered all around Russia [2]. If the two assets are now excessive for Russia, it would be wise to create favourable conditions for their use by the international scientific community for the

V A Koptyug and J Klerkx (eds), Science Policy New Mechanisms for Scientific Collaboration between East and West, 89–100

benefit of Russia and other countries. These premises seem to me much more reasonable than those mentioned above.

I am not a politician, but it seems to me very evident that a loss of respect for science and scientists, for higher education, for knowledge and wisdom in Russia is not only pitiful, but also dangerous to the entire civilized world. The results of the last election in the Russian parliament have confirmed my conviction.

Hence, I sincerely believe that NATO countries are interested in the survival of Russian fundamental science. However, they have many problems of their own, including social tension in some fields of basic research caused by lack of finances, and the only realistic way of supporting Russian science by NATO states is by mutually profitable and cost-efficient international scientific cooperation.

Since 1988, Lake Baikal the greatest, deepest and oldest lake in the world, a strategic deposit of 20% of the liquid fresh water of the Earth, home of 1500 endemic species of aquatic organisms, has become an object of large-scale international collaboration in the field of basic research. This move has been shaped by the Siberian Branch of the Russian Academy of Sciences into a kind of open institute — The Baikal International Centre for Ecological Research (BICER). Participants of BICER believe that it has become an extremely cost-efficient mechanism of interdisciplinary scientific cooperation with Russia, and propose to propagate its experience elsewhere by means of support on government and non-governmental levels. The high cost-efficiency of BICER is due to the ready availability of the two assets mentioned above — well-educated partners and diverse facilities — to scientists coming from abroad.

BICER was officially opened in December of 1990. Its founding Members are:

— Siberian Branch of the Russian Academy of Sciences;
— Royal Belgian Institute for Natural Sciences;
— Royal Society of London, Great Britain;
— Japanese Association for Baikal International Research Programs;
— Swiss Federal Institute of Technology;
— University of South Carolina, USA.

According to its Chapter, BICER organizes bilateral and multilateral joint research by Russian, as well as foreign, scientists along the following directions:

— multidisciplinary studies of the ecological system of Lake Baikal by methods of classical and physico-chemical biology, hydrochemistry, climatology, applied mathematics, oceanology and limnology, satellite and other methods of remote sensing;
— studies of the mechanism and chronology of the formation of the endemic biological species of Lake Baikal by methods of classical and physico-chemical biology, molecular genetics, paleolimnology and biogeochemistry as important parts of paleobiogeography;
— studies of global cycling of elements (carbon, nitrogen, sulphur, heavy metals, etc.) and of the global transfer of the most important eco-toxicants (polychlorobiphenyls, dioxins, radionuclides and other substances) with Baikal acting as a "back ground" site (remote from large-scale sources of pollution) by the most sensitive methods as instruments;
— multidisciplinary geological, paleogeographic, paleontological, geochemical studies of the history of Lake Baikal and its pre-existing water bodies in the Baikal Rift zone, as a scenario of geodynamic processes and of the formation of the unique biota with a special emphasis on the problem of Global Change;
— development of the scientific basis for integrated monitoring of the state of Lake Baikal.

BICER is based on the facilities of the Limnological Institute in Irkutsk. This Institute has a staff of 394 people, of them 251 are directly involved in scientific research, 3 doctors of science, 52 candidates of science working in laboratories of physical limnology, climatology, hydrochemistry, analytical chemistry, hydrobiology and systematics of aquatic organisms, ichthyology, pinnipeds biology, paleolimnology, molecular biology, electron microscopy. It possesses six research vessels of different size, of them the greatest R/V Vereshchagin with a displacement of 500 tons. Its main building has an area of 6000 m^2. It has a garage with expedition cars, a laboratory building and a few field laboratories on the shores of Lake Baikal. The Institute was opened in 1961 on the basis of the Baikal Limnological Station which had existed since the 1920-s. It now concentrates a considerable part of all the multidisciplinary scientific knowledge on Lake Baikal and acts as a major non-governmental consultant on matters concerning protection of the Lake.

— Other institutes of the Irkutsk Scientific Centre of the Siberian Branch of the Russian Academy of Sciences, like the Institute of the Earth's Crust, the Institute of Solar-Terrestial Physics, and the Institute of Geochemistry, take a very active part in BICER scientific activities and provide additional logistic support. The Institute of Geochemistry supervises (on the Russian side) one of the most important ventures of BICER — the Baikal Drilling Project (see the report of M.I. Kuzmin *et al.* [3] in this issue).

BICER, with Russian scientists, is willing to study Lake Baikal. BICER, according to its Chapter, is open to scientists of all countries who, together with Russian scientists, are willing to study Lake Baikal. The financing of BICER activities mainly comes with particular scientific projects, it does not have any permanent budget. This means that foreign participants come on their own money, obtained from national, and international science-funding agencies, bring (and take back) their own scientific instruments, and combine their efforts with the efforts of Russian colleagues who also take care of the financing of their projects themselves. Depending on particular scientific contributions, the results are published jointly, or separately. It has to be mentioned particularly that Russian partners (principal investigators) are not necessarily members of Limnological Institute, or of the Institutes of the Irkutsk Scientific Centre — they may come from any Russian institute or university, depending on scientific merit.

— As for BICER Founding Members, they agreed to take part in increasing facilities of BICER by allocation of special Founding Funds — money which is spent, on approval by consensus of the Founding Members Council for purchase of equipment, and other forms of institutional support. The minimal fee of a Founding Member is $ 100000. It is not immediately transferred to Russia, but rather remains in the country of a Founding Member before being used according to a common decision. If necessary, allocation of the Founding Fund may be done step-by-step; for example, BICER Founding Members Council agreed that Swiss Federal Institute of Technology will donate to BICER $ 20 000 every year in 1993—1997. Foreign contributions to the Founding Fund may be done in kind, in form of instruments, reagents, etc. Because of the financial crisis in Russia, part of the Founding Fund had to be spent, according to a decision of the Founding Members Council, for current needs, like purchase of fuel, repair of ships, etc. The Siberian Branch of the Russian Academy of Sciences, in spite of all the difficulties, gives BICER special financial support, due to which it acquired a Philips scanning electron microscope, personal computers, Applied Biosystems instruments for sequencing of nucleic acids and proteins, imported equipment for biochemical microanalysis, and received some

$ 150 000 for purchase of a tug which will be reconstructed into an ice-breaker for support of winter drilling operations.

Founding Members have the privilege of using all the facilities of BICER and establishing priorities, if these facilities, like ship time, are limited. All scientific, financial and logistic matters are settled, and plans are approved during BICER Founding Members Council meetings, which take place in December of every year.

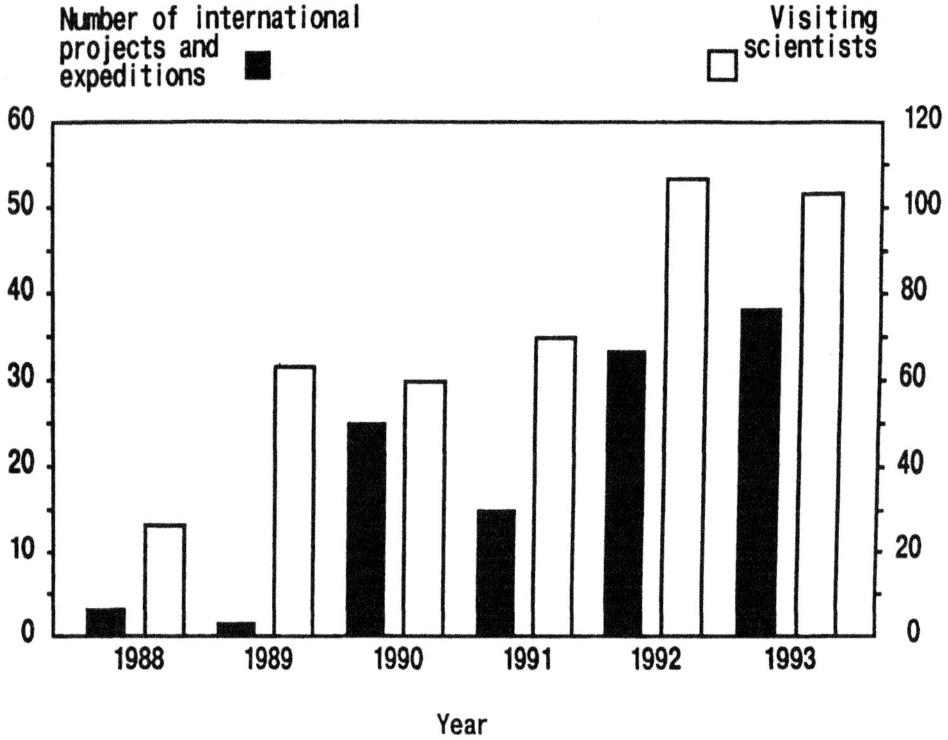

Fig. 1.
Number of foreign scientists accepted by BICER
and international projects of BICER in 1988—1993.

For the last few years, BICER has hosted many hundreds of foreign scientists (Fig. 1). At present we run some 40 international projects of different scale per year, which is close to the limits of our capacity. Participants come from Australia, Finland, Italy, Great Britain, New Zealand, the USA, Belgium, Japan, China, Switzerland, France, Germany, Sweden, Izrael, Canada, and from many cities of Russia and the former Soviet Union, like Moscow, Vladivostok, Krasnoyarsk, Ulan-Ude, Chita, Gelendgik, Minsk, St. Petersburg, Borok, Yaroslavl, Tartu, Novosibirsk, Barnaul, Odessa, Yakutsk, Baikalsk, Petrozavodsk and Irkutsk.

Fig. 2 shows data illustrating the participation of foreign scientists in BICER in 1993, expressed as man-days per country spent by them in international expeditions. I do not have the possibility of mentioning many special projects. Some of them, like joint use of Russian manned deep-water submersibles belonging to the Institute of Oceanology of the Russian Academy of Sciences (Moscow), very expensive and fruitful seismic profiling studies, and drilling of Baikal sediments, are briefly described in an accompanying paper [3]. The disciplines are diverse, from archaeology to analytical chemistry, from physical limnology to biological speciation, from molecular virology to estimating the standing stock of fish. The list of references attached (Addendum 1) gives the names of a majority of papers of BICER participants published in international journals, and in Russian journals by international groups. Many more papers are due to appear soon.

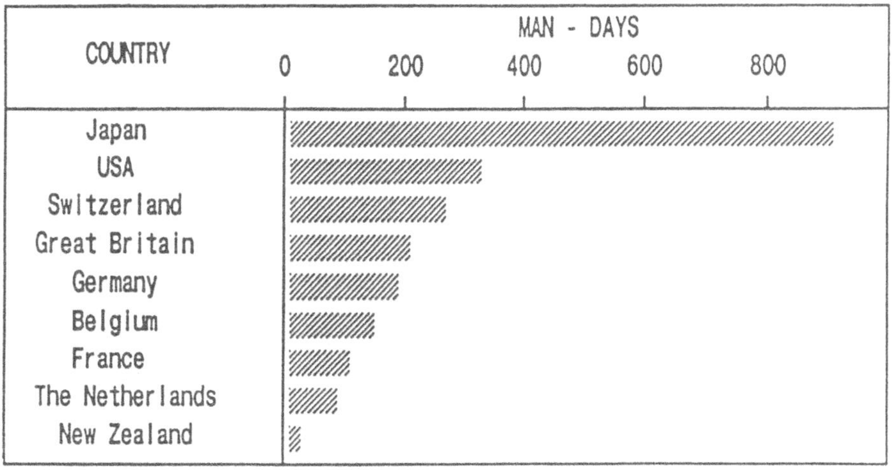

Fig. 2.
Participation of foreign scientists in BICER expeditions in 1993
(man-days per year per country).

I neither know how much money was spent by our foreign guests to come to Baikal, nor how much to bring and take many tons of expensive equipment, nor how much to deliver samples to their laboratories and to analyze them. Evidently, they would not have been able to come, if their projects had not been supported by national and international granting agencies. This means that their projects were good, that they had a global importance, and that they could not be done at home. Success in receiving grants, and in implementation of the projects depended to a great extent on participation of Russian scientists, who have studied Baikal for many decades and aquired a detailed knowledge of its ecosystem, geological history, endemic organisms, sources and the extent of pollution.

However, it is very simple to estimate our direct joint expenses in Russia. Fig. 3 shows total monthly expenses of the Limnological Institute, including those used for logistic support of all BICER projects, like salary of our staff and social taxes for 400 employees, cost of fuel and maintainance of six ships, per diem money and payment for hotel rooms of foreign and Russian guest scientists, maintainance of our main building and field laboratories, purchase of equipment and reagents, etc. Money actually spent was in rubles, but expenses are expressed in dollars per month according to the current free market course. Obviously there is hard currency inflation: to support all our activities, it was sufficient to have $ 5000 per month in the beginning of 1992, whereas in the end of 1993 we needed $ 60 000 per month. A major part of this money came from the Siberian Branch of the Russian Academy of Sciences. Foreign BICER members contributed some $ 50 000, non-member participants — some $ 20 000. This financial aid was very important for our success, because money came at those moments when financing from Russian sources did not come. One may easily imagine what would happen if our sailors or drivers of expedition cars would announce a strike, or if we had to send our scientists on a non-paid vacation, or if even during a short period we were not able to buy ship fuel (as happened in 1993 with many institutes in Russia), or we were forced to sell our ships — the Limnological Institute as a scientific unit could be lost forever, not to mention the ruined plans of our foreign partners.

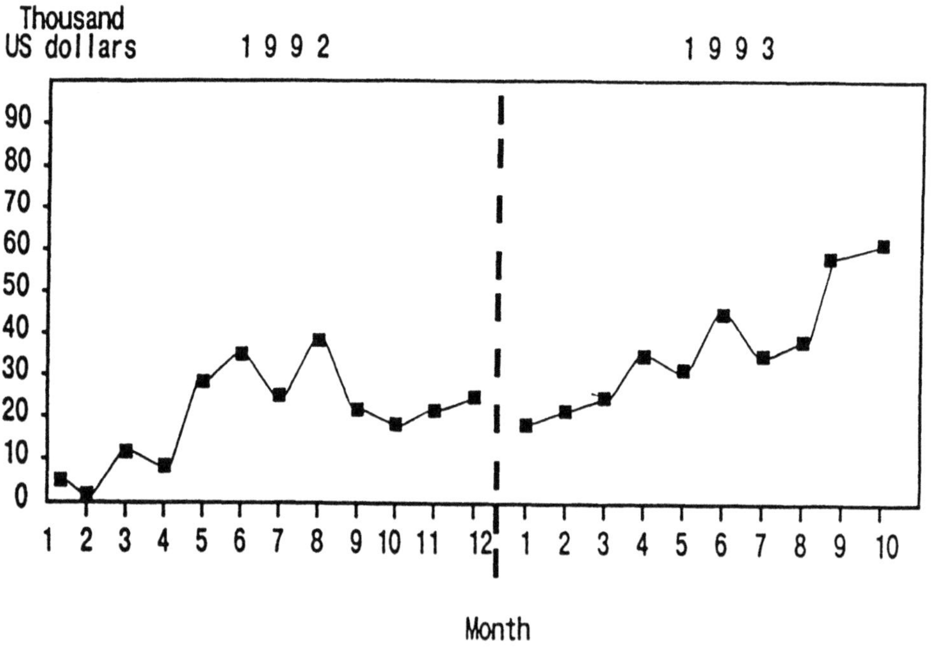

Fig. 3.
Monthly expenses of Limnological Institute in 1992-1993.

On the other hand, it is evident that the cost efficiency of the foreign support of Russian science through BICER was very high. It is orders of magnitude greater than that provided by other means, like the invitation of Russian scientists to foreign laboratories, purchase of journals, computers, and reagents, and even dedicated hard-currency grants to individual Russian scientists and research groups. From an economic viewpoint, this high cost efficiency is not surprising, because current support, like a catalyst, helped to activate huge capital — an interdisciplinary team of highly qualified Russian scientists, and a very expensive infrastructure — for the benefit of international science.

References

[1] Editorial (1993). Nature v. 365, pp. 475—476.
[2] Programme of Development of International Research Centres in Siberia, Siberian Branch of the USSR Academy of Sciences, Sibvneshtorgreklama (1991). Novosibirsk.
[3] Kuzmin M.I. *et al.*, this issue.

♦

Addendum 1

GENERAL BIOLOGY

Evstigneeva T., Clark P.F., Boxshall G.A. Nocturnal emergence behaviour of the fauna of Lake Baikal, Siberia.
J. Prog. Underwater Sci.; 1991; 16: 85-88.

Timoshkin O.A. Turbellaria Lecithoepitheliata: morphology, systematics, phylogeny.
Hydrobiologia; 1991; 227: 323-332.

Martens K., Mazepova G. On Limnocythere baikalensis n. sp. from Lake Baikal (Siberia, USSR), with notes on the position of the L.goersbachensis-group (Crustacea, Ostracoda, Limnocytheridae).
Arch. Hydrobiologie (Monographishe Beitrage); 1992; Suppl.90; 1: 115-131.

Martens K., Noskova I., Mazepova G. On Pseudocandona ceratina Mazepova.
Stereo-Atlas of Ostracod Shells; 1992; 19(11): 41-48.

Martens K., Noskova I., Mazepova G. On Pseudocandona gajewskajae Bronstein
Stereo-Atlas of Ostracod Shells; 1992; 19(12): 49-56.

Likhoshway Ye.V., Yakushin A.O., Puzyr A.P., Bondarenko N.A. Fine strucstructure of the velum and girdle bands in Aulacoseira baicalensis.
Diatom Research; 1992; 7(1): 87-94.

Karpov S.A., Tanichev A.I. The Ultrastructural Study of Green Alga Pedinomonas tenuis Masiuk, 1970 with Special Reference to the Flagellar Apparatus.
Arch. Protistenkd.; 1992; 141: 315-326.

Flower R.J. A taxonomic re-evaluation of endemic Cyclotella taxa in Lake Baikal, Siberia.
Nova Hedwigia, Beiheft; 1993; 106: 203-220.

Boxshall G.A., Evstigneeva T.D., Clark P.F. A new interstitial cyclopoid copepod from a sandy beach on the western shore of Lake Baikal, Siberia.
Hydrobiologia; 1993; 268: 99-107.

Jakubowski M. Morphometry of the Gill Respiratory Area in Comephorus dybowskii and Some Other Endemic Cottoidei of Lake Baikal.
Acta Zoologica (Stockholm); 1993; 74(4): 283-288.

Kamaltynov R.M., V.I.Chernykh, Slugina Z.V., Karabanov E.B. The consortium of the sponge Lubomirskia baicalensis in Lake Baikal, East Siberia.
Hidrobiologia, 1993, 271: 179-189.

Addendum 1 (cont.)

MOLECULAR BIOLOGY AND BIOCHEMISTRY

Vysotskii M V , Imbs A B , Popkov A A , Latyshev N A , Svetashev V I
Trans-olefinic very-long-chain fatty acid (26·3 4c,9c,19t) in lipids of fresh-water sponges of Lake Baikal.
Tetrahedron Letters; 1990; 31(30): 4367-4370.

Baram G I , Grachev M A , Malikov N G , Nazimov I V , Shemyakin V V
Amino Acid Sequence of Baikal Seal's Myoglobin.
Bioorg.Khimia; 1991, 17: 1166-1171

Grachev M A , Slobodyanyuk S Ja , Kholodilov N G , Fyodorov S P , Belikov S I , Sherbakov D Yu , Sideleva V G , Zubin A A , Kharchenko V V
Comparative Study of Two Protein-Coding Regions of Mitochondrial DNA from Three Endemic Sculpins (Cottoidei) of Lake Baikal.
J. Mol. Evol.; 1992; 34: 85-90.

Fedorov A N , Grechko V V , Slobodyanuk S Ya , Fedorova L V , Timokhina G I Taxonomic analysis of DNA repeats.
Mol.biologia; 1992, 26: 464-469.

Green J , Kozlova T Carotenoids, photoprotection and food web links in Lake Baikal.
Freshwater Biology; 1992; 28: 49-58.

Ivanova Ye P , Bakunina I Yu , Gorshkova N M , Romanenko L A , Michailov V V , Elyakova L A , Parfenova V V Distribution of chitine-degrading enzymes in marine and frech-water microorganisms
Biologia Morya (Marine Biol., Russ.); 1992, 22: 464-469.

Eriksson K , Reuter M , Timoshkin O Analysis of catecholamines in Turbellarians of Lake Baikal.
Comp. Biochem. Physiol.; 1993; 105(3): 509-511.

Trofimova I N , Kuznedelov K D , Kumarev V P Molecular cloning and sequencing of omul prolactin cDNA.
Molecular Marine Biology and Biotechnology; 1993, 2(1): 41-43.

ENVIRONMENT POLLUTION

Oikari A O , Baram G I , Evstafyev V K , Grachev M A Determination and characterization of chloroguaiacol conjugates in fish bile by HPLC
Environmental Pollution; 1988; 55: 79-87.

Maatela P , Paasivirta J , Grachev M A , Karabanov E B Organic chlorine compounds in lake sediments V Bottom of Baikal near a pulp mill.
Chemosphere; 1990; 21(12): 1381-1334.

Kucklick J R , McConnell L L , Bidleman T F , Ivanov G P , Walla M D
Toxaphene contamination in Lake Baikal's water and food web
Chemosphere; 1993; 27(10): 2017-26.

Addendum 1 (cont.)

PALEOLIMNOLOGY

Bezrukova Ye V ,Bogdanov Yu A ,Williams D F ,Granina L Z , Grachev M A , Ignatova N V , Karabanov E B , Kuptsov V M , Kurilev A V , Letunova P P , Likhoshway Ye V , Chernyaeva G P , Shimaraeva M K A dramatic change of the ecosystem of Lake Baikal in Holozene.
Dokl.Akad.nauk SSSR; 1991, 321:1032-1036.

Likhoshway Ye V , Yakushin A O , Puzyr A P , Bondarenko N A Fine structure of the velum and girdle bands in Aulacoseira baicalensis.
Diatom Research; 1992; 7(1): 87-94.

Colman S M , Williams D F , Hearn P P Yr , Grachev M , Kuzmin M , Khahkaev B N Lake Baikal paleoclimate members. Initial results of U.S.-Soviet Paleoclimate study of Lake Baikal.
Eos; 1992; 73(43): 457-463.

Colman S M , Kuptsov V M , Johns G A , Carter S J Radiocarbon dating of Baikalian sediments.
Geologiya i geofizika; 1993; 34(10-11): 68-77.

Deike R , Webster D , Callender E Sedimentary changes and authigenic mineralogy, cores from site 307P, 307A-3 Academician Ridge, Lake Baikal.
Geologiya i geofizika; 1993; 34(10-11): 160-173.

King J W , Pek J , Ganguemi P , Kravchinskii V A Paleomagnetic and petromagnetic studies of sediments of Lake Baikal.
Geologiya i geofizika; 1993; 34(10-11): 174-191.

Kuzmin M I , Williams D F , Logachev N A , Colman S , Khakhaev B N , KawaiT , Hern P , Horie Sh ,Pevzner L A , Bukharov A A , Fialkov V A The Baikal drilling program: tasks, proceeding, solving process problems
Geologiya i geofizika; 1993; 34(10-11): 5-15.

Orem W H , Lerch H E , Kotra R K Products of lignin oxidation in Lake Baikal sediments, Russia: indicators of late-quaternary paleovegetation and paleoclimatic changes in Northern and Central Asia.
Geologiya i geofizika; 1993; 34(10-11): 108-123.

Spiker E C , Bates A L Sulfur isotope effects of Late Quaternary climate change in Lake Baikal
Geologiya i geofizika; 1993; 34(10-11): 102-107.

Toyoda K , Takemura K , Ishiwatari R Complex study of the 323-PC1 core of the Baikal Lake bottom sediments
Geologiya i geofizika; 1993; 34(10-11): 124-135.

Addendum 1 (cont.)

GEOPHYSICS

Hutchinson D R , Golmshtok A J , Zonenshain L P , Moore T C , Sholz C A , Klitgord K D Preliminary results of multichannel seismic reflection surveg carried out in Baikal Lake in 1989.

Geologiya i geofizika; 1993; 34(10-11): 25-36.

Scholz C A , Klitgord K D , Hutchinson D R , Ten Brink Uri S , Zonenshain L P , Golmshtok A Y , Moore T C Results of 1992 seismic reflection experiment in Lake Baikal.

Eos; 1993; 74(41): 469-470.

PHYSICAL AND GENERAL LIMNOLOGY

Back R C , Bolgrien D W , Guselnikova N E , Bondarenko N A Phytoplankton photosynthesis in Southern Lake Baikal: size-fractionated chlorophyll a and photosynthetic parameters.

J. Great Lakes Res.; 1991; 17(2): 194-202.

Grachev M A Slow renewal of deep waters.

Nature; 1991; 349: 654-655.

Falkner K K , Measures Ch I , Herbelin S E , Edmond J M The major and minor element geochemistry of Lake Baikal.

Limn.and Oceanogr.; 1991; 36(3): 413-423.

Edgington D N , Klump J V , Robbins J A , Kusner Y S , Pampura V D , Sandimirov I V Sedimentation rates, residence times and radionuclide inventories in Lake Baikal from Cs-137 and Pb-210 in sediment cores.

Nature; 1991; 350: 601-604.

Weiss R F ,Carmack E C , Koropalov V M Deep- water renewal and biologikal production in Lake Baikal

Nature; 1991;349(6311): 665-669.

Rudstam L G , Melnik N G , Timoshkin O A , Hansson S , Pushkin S V , Nemov V Diel dynamics of an aggregation of Macrohectopus branickii (Dyb) (Amphipoda, Gammaridae) in the Barguzin bay, Lake Baikal, Russia

J. Great Lakes Res.; 1992; 18(2): 286-297.

Shimaraev M N , Granin N G , Zhdanov A A Deep ventilation of Lake Baikal waters due to spring thermal bars

Limnol.Oceanogr.; 1993, 38(5):1068-1072

Addendum 1 (cont.)

THERMAL VENTS

Crane K , Hecker B , Golubev V Hydrothermal vents in Lake Baikal
Nature; 1991; 350: 281.

Kuznetsov A P , Strizhov V P , Kuzin V S , Fialkov V A , Yastrebov V S
News on nature of Baikal. A communite based upon bacterial chemosynthesis.
Izv.Akad.nauk SSSR, ser.biol.; 1991, 5: 766-772

Shanks III W C , Callender E Thermal springs in Lake Baikal.
Geology; 1992; 20: 495-497.

Sitnikova T Ya , Fialkov V A , Starobogatov Ya I Gastropoda from underwater hydrothermal vent of Baikal Lake.
Ruthenica; 1993; 3(2): 133-136.

VIROLOGY

Grachev M A , Kumarev V P , Mamaev L V , Zorin V L , Baranova L V , Denikina N N , Belikov S I , Petrov E A et al Distemper virus in Baikal seals.
Nature; 1989; 338: 209.

Likhoshway Ye V , Grachev M A , Kumarev V P , Solodun Yu V , Goldberg O A , Belykh O I , Nagieva F G , Nikulina V G , Kolesnik V S Baikal seal virus.
Nature; 1989; 339: 266.

Osterhaus A D M E , Groen J , Uytdehaag F G C M , Kumarev V P , Zorin V L Different Morbilliviruses in Europian and Sibirian Seals.
Vet. Rec.; 1989; 125: 647-648.

Barrett T , Visser I K G , Mamaev L , Goatley L , van Bressem M F , Osterhaus D M E Dolphin and porpoise mobilliviruses are genetically distinct from phocine distemper.
Virology; 1993; 193.

AEROSOLS

Bashurova V S , Dreiling V , Hodger T V , Jaenicke R , Koutsenogii K P , Koutsenogii P K , Kraemer M , Makarov V I , Obolkin V A , Potjomkin V L Pusep A Y Measurements of atmospheric condensation nuclei size distribution in Siberia.
J. Aerosol Sci.; 1992; 23(2): 191-199.

Koutsenogii P K , Bufetov N S , Drosdova V I , Golobokova L P , Khodger T V , Kutsenogii K P , Makarov V I , Obolkin V A , Potemkin V L Ion composition of atmospheric aerosol near Lake Baikal
Atmospheric Environment; 1993; 27A(11): 1629-1633.

A BRIEF REPORT OF JAPANESE ACTIVITY FOR BICER PROJECTS IN 1993

T. KAWAI
National Institute for the Environmental Studies,
Japanese Association for the Baikal International Research Program,
Tsukuba, Japan

ABSTRACT. The purpose of this report is to present a brief description of the activities of the Japanese Association for the Baikal International Research Program (JABIRP) in 1991-1993.

Keywords: lake Baikal / ecosystems / coordinator / funding contribution.

Introduction

The establishment of the Baikal International Center for Ecological Research (BICER) was recognized by many Japanese scientists as being timely for addressing many global environmental problems. Lake Baikal and its watershed can provide us with excellent samples in researching the reconstruction of paleolimnological change of this area, the taxonomy and evolution of endemic species, the ecological system of a large non-disturbed watershed, and so on. The opening of the Baikal area to scientists throughout the world is thought of as one of Russia's greatest contributions (the former USSR) in the last decade of the 20th century, and will be a tribute to the Russian people in future.

The purpose of this report is to present a brief description of the activities of the Japanese Association for the Baikal International Research Program (JABIRP) to the International Advanced Research Workshop "Science Policy: New Mechanisms for Scientific Collaboration Between East and West" (Novosibirsk, Nov. 22-25, 1993).

1. History of JABIRP

The plan for BICER was introduced to Japanese scientists in December 1989, at first via the office of the Ministry of Foreign Affairs and the Japanese Environmental Agency. Dr. M. Aizaki, from the National Institute for Environmental Studies, was sent as an observer to the international workshop organizing BICER, which was held in May 1990. After receiving the report of Dr. Aizaki, members from the National Institute for Environmental Studies, the Environmental Agency, the Japanese Association of Limnology, and the Japanese Association of Ecology gathered to discuss BICER. This group inquired as to how many Japanese scientists were interested in joining the international collaboration in the Baikal area. They had received many positive replies and requests from Japanese colleagues by the end of December 1990.

V. A. Koptyug and J. Klerkx (eds.), Science Policy: New Mechanisms for Scientific Collaboration between East and West, 101–106.

Dr. Y. Tomoda, who attended the international meeting in December 1990 in Irkutsk at which the construction of BICER was discussed, informed us that the American and Belgian participants had agreed to provide a fundamental research fund to start BICER and that the coordinating committee had declared BICER open. In late January 1991, the voluntary members agreed to establish the Japanese Association for the Baikal International Research Program (JABIRP) by the end of March 1991. They elected Dr. T. Kawai from the National Institute for Environmental Studies as the secretary of the organizing committee. JABIRP was established on March 30, 1991 by 33 Japanese scientists. Prof. Okuda was elected director.

Events following JABIRP's Establishment:

1991	March 30	Start of JABIRP with 33 members.
	April 27	Meeting with Prof. Grachev (Tokyo).
	June	Completed arrangement for 100000 USD, and applied for membership on the BICER FMC.
	August	First visit to BICER (Irkutsk) and Expedition to Lake Baikal (10 members).
	December	BICER FMC (Prof. Numachi, Prof. Wada, Dr. Kawai). Total 12 members visited Irkutsk during 1991.
1992	March	JABIRP becomes the third member of BDP.
	April	The 2nd Annual Meeting of JABIRP (68 members).
	May	Two JABIRP groups received research funds from the Ministry of Education.
	August 19	Dr. Karabanov was invited to visit 4 cities and joint IGC.
	December	BICER FMC (Dr. Kawai, Prof. Miyazaki). Total 10 JABIRP teams including 37 members visited Irkutsk and Lake Baikal in 1992. A Fundamental contribution of 101400 USD was completed.
1993	January 25	Start of Baikal Sen-nin Kai (Personal Sponsors Organization).
	February 22	Japanese-Russian Treaty for Science and Technology Cooperation. Two themes relating to BICER were listed.
	April 26	The 3rd Annual Meeting of JABIRP (106 members).
	May 26	Special Report of the Japan Science Council about the significance of and expectations for BICER.
	September 15	Delegation of the Japan Science Council for the development of collaboration visits by BICER and Science Center of Irkutsk Branch, SB RAS.
	December 10	BICER FMC (Prof. Wada, Dr. Kawai). Total 14 JABIRP teams including 46 members and 11 scientific tourists in 1993, and the JSC Delegation.

2. Organization of JABIRP

As of November 1, 1993 JABIRP had 115 members from throughout Japan. They are now working on many subjects under BICER, either belonging to or independent from the main programs.

(1) JABIRP GENERAL COORDINATING STAFF

Organizing staff help members and research groups by arranging common contributions of JABIRP to BICER, distributing general and international information, supporting intergroup communication, and playing a role as the formal representative of JABIRP.

The staff include:

Honorary Director: Dr. Jiro Kondo, President of the Japanese Science Council.

Advisory Board: Drs. Tatsuo Kira, Tetsuya Torii, Koji Fushimi.

Director: Prof. Setsuo Okuda(*), Faculty of Science, Okayama University of Science.

Vice-Director: Prof. Hiroya(*) Kawanabe, Director of the Center of Ecological Research, Kyoto University.

Representatives to the BICER FMC: Prof. Ken-ichi(*) Numachi, Faculty of Oceanology, Tokai University; Prof. Eitaro Wada(*), CER, Kyoto University.

Scientific Advisory Board of BICER: Prof Kawanabe, Dr. Morino.

Head Research Coordinator: Prof. Eitaro Wada(*).

Vice-research Coordinator: Dr. Akira Goto, Hokkaido University, staff: 10 Drs. (mainly chiefs of the research groups).

Workshop for Future Projects: Drs. Y. Nojiri, T. Yoshioka, staff: 8 Drs. (younger leaders).

Editorial Board: Drs. Hiroshi Morino, Nobuyuki Miyazaki, preparing a book "Lake Baikal" to be published in Japanese in May 1994 editor of the JABIRP Annual Report.

Secretary General: Dr. Takayoshi Kawai(*), National Institute of Environmental Studies.

Vice-secretary: Prof. Nobuyuki Miyazaki, Tokyo University, 10 secretaries.

International Research Communication: Dr. Hiroshi Morino.

Representative of the Baikal Research Foundation (JABIRP): Mr. Hiroshi Okazaki, President of the Global Environmental Forum (Japan).

Baikal Sen-nin Kai: Dr. Takayoshi Kawai (secretary), 11 sponsors.

The Higashi-mikawa Branch: Mr. Tatsuzo Ban (Head), 33 sponsors.

(2) JABIRP RESEARCH GROUPS

Each research group works largely independently from the JABIRP coordinating staff, for example, organizing collaborators, selecting counterparts, arranging their research funds, publishing the results and so on. Representative organizers of each group are responsible for almost all of their research activities. Communication with coordinating staff is easily maintained because most representative organizers also work as the coordinating staff.

The Head research coordinator in the general coordinating staff, however, sometimes conducts discussions about significant subjects to be supported or initiated from the view point of total development of the project, selects a suitable representative organizer for a new subject, and helps him/her to obtain research funds.

The three projects mentioned in section 3 are being developed as components of the main programs of JABIRP under the support of the coordinating staff.

(*) summit member of JABIRP

3. JABIRP Scientific Objectives and Research Program

JABIRP is arranging three main programs following the philosophy of the BICER charter, and supporting various subjects of small teams and personnel.

(1) PRESENT ECOSYSTEM IN LAKE BAIKAL AND ITS CATCHMENT-AREA

Understanding biological-geological-climatical interactions and scientific support for effort to protect the ecosystem.

Prof. Setsuo Okuda(*): Representative organizer,
Prof. Eitaro Wada: Vice-organizer,
Dr. Setsuo Okuda: chief of physical limnology,
Dr. Yasuhiro Watanabe: chief of phytoplankton ecology,
Dr. Akira Takeuchi: chief of geological survey, (supported by the Ministry of Cultural, Science and Education).

(2) TAXONOMY AND EVOLUTION OF ENDEMIC SPECIES IN LAKE BAIKAL: ROOTS OF THE BAIKAL SEAL AND EVOLUTIONARY HISTORY OF GAMMARIDS AND BULLHEADS

Prof. Ken-ichi Numachi(*): Representative organizer,
Prof. Nobuyuki Miyazaki: Vice-organizer,
Dr. Akira Goto: chief of fish group,
Dr. Hiroshi Morino: chief of crustacean group,
Dr. Shinsuke Tanabe: chief of pollution group, (supported by the Ministry of Culture, Science and Education).

(3) BAIKAL DRILLING PROJECT: PALEOLIMNOLOGICAL RECONSTRUCTION OF LONG-TERM ECOLOGICAL, CLIMATICAL AND GEOLOGICAL CHANGE OF LAKE BAIKAL AND ITS BASIN

Dr. Takayoshi Kawai(*): Representative organizer of the BDP (JABIRP),
Dr. Yoshio Inouchi: Vice-organizer,
Dr. Genki Matsumoto: chief of the chemical group,
Prof. Kenji Kashiwaya: chief of physics and sedimentology,
Dr. Kimiyasu Kawamuro: chief of biological information,
Dr. Akira Takeuchi: chief of the dating group,
Dr. Masayuki Kunugi: chief of the BDP data base, (applying to the Science and Technology Agency).

(4) SUBJECTS RESEARCHED BY SMALL TEAMS OR INDIVIDUALS

4. Founding activity in Japan

(1) FOUNDING CONTRIBUTION OF JABIRP FOR SUPPORTING BICER ACTIVITY

The total contribution for three years since the 1991 is 177900 USD. The contribution of JABIRP to BICER is not yet sufficient to realize the potential of the Baikal Programs. Expansion of scientific activity will enlarge the contribution.

1991: total 6600 USD (1 USD=130 JYN)
1992: total 127900 USD including FMC fund (1 USD=125 JYN),
    101700 USD, fundamental contribution as an FMC member,
    25000 USD, contribution of JABIRP to the Baikal Drilling Project,
    1200 USD, consumables, limited to the counterparts, etc.
1993: total 45400 USD (1 USD=110 JYN)
    20000 USD, fundamental contribution as an FMC member,
    20000 USD, JABIRP contribution to BDP,
    5400 USD, consumables, limited to the counterparts, etc.

(2) FOUNDING ACTIVITY IN JAPAN

Since inception, JABIRP has been making efforts to procure research funds. There are essentially four types of funding available to support fundamental research activity in Japan. The first two are governmental, the third is a non-profit organization and the forth source is from private companies or personal. Support of Scientific tours to Lake Baikal also provides an income for research.

1) From the annual budget of institutes or universities all JABIRP members, except students, receive a small fundamental research grant and a full salary from his/her institute or university. For example, in 1993, Dr. Kawai received a supplemental grant of 2500000 JYN (22000 USD) to support the Baikal program.

2) From the special foundation of each Ministry or Agency. In 1993, three groups received a total of 18000000 JYN (160000 USD) from the Ministry of Culture, Science and Education. In the future, other groups will apply to receive more funds from other Ministries or Agencies such as the Japanese Science and Technology Agency (STA), the Environment Agency, etc.

3) From licensed foundations. The Global Environmental Forum (President H. Okazaki) has been assisting, since the inception of JABIRP, in the arrangement of personal contributions, and sometimes in procuring funds from private foundations. The forum has a special account for funding the Baikal research programs. JABIRP gets about 6000000 JYN (54000 USD) from private sponsors and foundations through this account.

4) From private foundations and sponsors JABIRP is partly supported by the private funding organization named "Baikal Sen-nin Kai" which is a special group of individual sponsors for the BICER activity in Japan. "Sen-nin" is an imaginary old man, a symbol of immortal, well-managed power. He is benevolent to the earth and of good nature. Each sponsor offers at least 30000 JYN (one unit of contribution) per year. JABIRP also asks sponsors to offer temporal contributions, which are sometimes of larger.

5. Expectations of BICER

The new challenge to international collaboration with BICER has excited many Japanese scientists. More and more colleagues will come to work together in Lake Baikal and its catchment-area for the rapid progress of basic field science and for the solution of global environmental problems. JABIRP participants expect continuous and stable development of BICER. JABIRP would like to appeal especially to Asian and Australian colleagues to attend BICER. Baikal could possibly be the world center of field science in the 21st century.

Appendix

List of JABIRP Visitors during 1993

Participants of Research Expeditions and meeting:

Takayoshi Kawai: 3 - 24 Feb.
Genki Inoue: 31 Mar. - 7 Apr.
Akio Misaki: 12 - 26 May
Shuichi Endo: 2 - 16 June
Ryohei Tsuda: 2 - 16 June
Yasushi Ichimura: 2 - 16 June
Akira Takeuchi: 23 June - 14 July
Yoshiaki Kikuchi: 7 July - 4 Aug.
Hidetoshi Naganawa: 7 July - 25 Aug.
Toshio Nakamura: 14 July - 12 Aug.
Shin-ichi Ishiwata: 14 July - 18 Aug.
Kouchi Yoshii: 14 July - 18 Aug.
Yasunori Watanabe: 28 July - 12 Aug.
Noboru Fujita: 11 - 25 Aug.
Hiroto Maeda: 12 - 17 Aug.
Takashi Sato: 25 Aug. - 1 Sept.
Mamoru Yabe: 8 Sept. - 23 Sept.
Akira Goto: 8 - 23 Sept.
Takayuki Hanazato: 29 Sept. - 13 Oct.
Ken-ichi Nimachi: 25 Oct. - 4 Nov.
Nobuyuki Miyazaki: 29 Oct. - 17 nov.
Haruhiko Nakada: 29 Oct. - 17 Nov.
Takayoshi Kawai: 8 - 15 Dec.
Shoji Fujii: 17 - 24 Feb.
Susumu Hayashi: 31 Mar. - 14 Apr.
Setsuo Okuda: 2 - 16 June
Yasuaki Okumura: 2 - 16 June
Koji Yokoyama: 2 - 16 June
Takeshi Matsubara: 16 June - 14 July
Shoji Fujii: 7 July - 12 Aug.
Akifumi Otaka: 7 July - 4 Aug.
Takayoshi Kawai: 14 July - 12 Aug.
Hiroshi Morino: 14 July - 25 Aug.
Kazuo Mashiko: 14 July - 4 Aug.
Yoshiki Masuda: 21 - 28 July
Toshiya Katano: 28 July - 12 Aug.
Susumu Hayashi: 11 Aug. - 15 Sept.
Ken-ichi Numachi: 18 - 23 Aug.
Shuzo Oi: 25 Aug. - 23 Sept.
Mutsumi Nishida: 8 - 23 Sept.
Takayoshi Kawai: 13 Sept. - 1 Oct.
Takahito Yoshioko: 29 Sept. - 13 Oct.
Hiroyuki Sasaki: 29 Oct. - 17 Nov.
Masao Amano: 29 Oct. - 17 Nov.
Masako Numachi: 18 Oct. - 1 Nov.
Eitaro Wada: 8 - 22 Dec.

Toyama TV: 15—23 Sept. Ichiro Maeda (reporter), Koichi Oyama (cameraman).

Prof. Okuda's Group for Scientific Tours, 15—23 Sept.

Setsuo Okuda, Yusaku Nogami, Tamiya Kishimoto, Mitsuo Tanaka, Kazutaka Nagao, Minoru Fukuhara, Shogo Ikeda, Shohei Kira, Makiko Hirata, Takahiro Sugita, Iwao Okamoto

Delegation of the Japan Science Council, 15 -18 Sept.

Jiro Kondo (president), Yasuo Okamoto, Yuiti Ono, Kozo Amano, Akira Okada, Yasuhiro Oda

♦

BAIKAL DRILLING PROJECT: FIRST RESULTS AND PROSPECTS FOR FUTURE STUDIES

M.I. KUZMIN
Vinogradov Institute of Geochemistry of SB RAS, Irkutsk, Russia

N.A. LOGACHEV
International Centre for Active Tectonics and Natural Disasters, Institute of the Earth's Crust of SB RAS, Irkutsk, Russia

M.A. GRACHEV
Baikal International Centre for Ecological Research, Limnological Institute of SB RAS, Irkutsk, Russia

P.P. HEARN
U.S. Geological Survey National Centre, Reston, Virginia, USA

D.F. WILLIAMS
University of South Carolina, Columbia, USA

Sh. HORIE
Kyoto University, Japan

T. KAWAI
National Institute for the Environmental Studies, Japanese Association for the Baikal International Research Program, Tsukuba, Japan

ABSTRACT. The organization and basic results of works in terms of the international program "Baikal Drilling" accomplished by the American, Russian and Japanese scientists are represented. The project is aimed at the study of global changes of environment and climate in Central Asia based on comprehensive investigations of Baikal sediments. The geophysical and geological works preceding drilling on Baikal allowed to establish the thickness (up to 7 km) and structure of the Baikal sedimentary basin, sedimentation rate, geochemistry of the upper sedimentary layer. The paper also yields the results of drilling the first 100 m borehole in 1993 which showed that sediments of Baikal may be used for interpretation of paleoclimate. The investigations are proposed to be continued in terms of the International Program "Mega Science Forum".

Keywords: paleoclimatic / rift zone / drilling / lake Baikal.

V A Koptyug and J Klerkx (eds), Science Policy New Mechanisms for Scientific Collaboration between East and West, 107–116

The present contribution is given on behalf of the Russian, American and Japanese scientists, who have been involved in investigations on paleoclimatic history of Lake Baikal for over four years.

Lake Baikal occupies the largest basin in the Baikal Rift zone, the one of two major continental rift zones in the world. Lake Baikal is located in Central Asia and consists of three basins: Southern, Central and Northern (Fig. 1). No other lakes similar to Lake Baikal are available at the same latitude in the northern hemisphere. Baikal contains a continuous sedimentary record that spans the last twenty to thirty million years, as the lake basin was not glaciated during the Pleistocene ice ages. Baikal sediments therefore ideally suited to reveal the record of geological and climatic changes in Central Asia including the growth of the young mountains, the uplift of the Tibet Plateau, and uplift of the Asian continent, which affect airflow circulation of the northern hemisphere and continental climate.

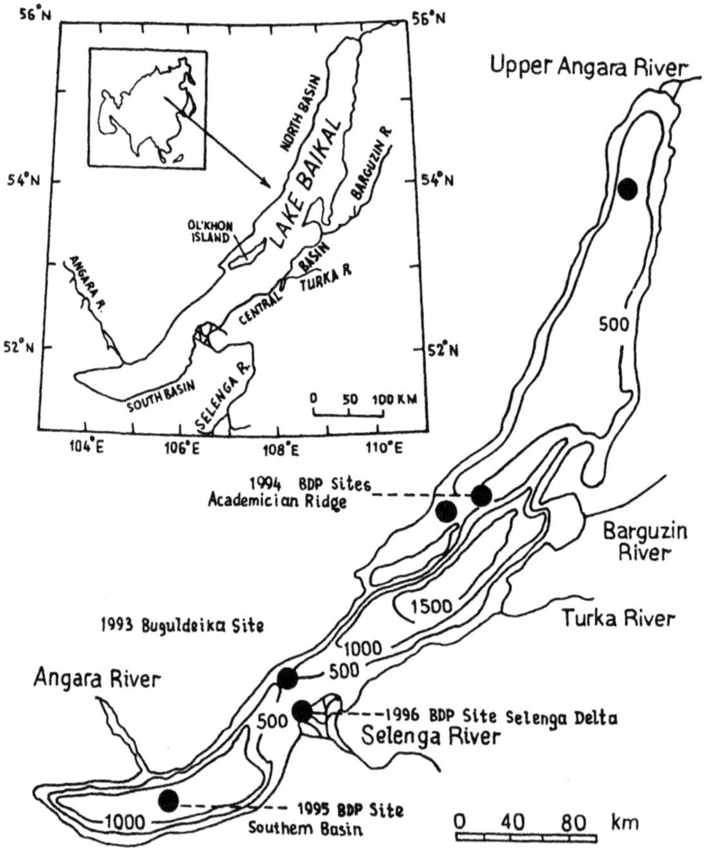

Fig. 1.
Schematic bathymetric map of Lake Baikal.
Black dots denote drill sites at Buguldeika and elesewhere.

In 1989 the team of the Russian and American and Japanese scientists started the program entitled "Baikal Drilling Project" (BDP) to study natural processes influencing the sedimentary record of Lake Baikal [3]. The project has become an important part of activities of the Baikal International Centre for Ecological Research (BICER) now uniting members represented by the Siberian Branch of the Russian Academy of Sciences, University of South Carolina, Japanese Association for the Baikal International Research Program (JABIRP) and a number of European countries.

The main goals of the Baikal Drilling Project are:

— to investigate the global change of the environment and climate in Central Asia based on the study of sedimentary record of Baikal;

— to investigate the evolution of sedimentary basins forming in the active continental rift zone.

To fulfill and coordinate the work of the Baikal Drilling Project a Steering or Coordination Committee, which includes Dr. M. Kuzmin (Russia), Prof. D. Williams (USA), Dr. T. Kawai (Japan), and a few workshops has been organized.

Several financial sources have been applied to make the project work successfully. In 1992—1993, the total contribution amounted to 100 mln roubles without including salaries of people engaged in the Project, that is purely financing of the drilling operations and related work only.

It was necessary to conduct geophysical and geological works to prepare the drilling. In 1989, the scientists from the Southern Division of Oceanology Institute executed over 1600 km of multichannel seismic survey. The data were processed jointly with American geophysicists, and resultant data are represented at the map of sedimentary thicknesses of the Lake Baikal bottom [1, 2]. The thickness of the bottom sediments is maximum on the western shore filling several deep basins (7—7.5 km deep).

The well stratified sedimentary cross-section through the Academic Ridge dividing Lake Baikal into the Central and Northern basins displays that it consists of three sequences (Fig. 2). Geophysicists distinguish (i) proto-rift (seismically transparent), (ii) mid-rift and (iii) modern rift sediments. The mid-rift sediments show pronounced stratification and evidence of intensive tectonic dislocations [1].

The Northern basin is markedly different from other basins structure and thickness of the sedimentary cover. There is no evident division into the proto-, mid- and modern rift deposits, and the thickness of sediments is not in excess of 3 km. The submersible investigations conducted on Lake Baikal in 1991—1992 contribute to a clearer understanding of its geology. Particularly important were underwater investigations performed in the area of the Academic Ridge, where the basement contacts overlapping sediments. Fig. 3 provides cross-sections through the underwater Academic Ridge, obtained via submersible studies. The basement is composed of Precambrian weathered rocks overlapped by the basal horizon of pebbles having beach origin, its age supposedly Upper Miocene. Above the cross-section there are cavernous clays of presumably Pliocene age. Thus, up until Upper Miocene the Academic Ridge, dividing the Central and Northern basins of Baikal, there was the land. The Northern basin has a young history, its development is related to the Baikal rift propagation northwards.

The Baikal sediments were the object of investigations before drilling operations were commenced. The sediments were sampled by piston and box coring nearly over the entire bottom of Baikal. The sediments were analyzed to determine their lithology, physical properties, mineral, chemical and rare element compositions. The results of these investigations were published in different Russian and International journals. Prof. Sh. Horie prepared a special issue of IPPCCE Newsletter №6, 1992. Dr. M.I. Kuzmin and Prof. D. Williams were in charge of the special issue in the Journal "Geology and Geophysics" №10 and №11 of 1993 published in English and Russian. The issue consists of twenty one articles of the Russian, American and Japanese scientists holding information on research preceding accomplishment of drilling, development of the project and its relevance to other investigations conducted in the Baikal region.

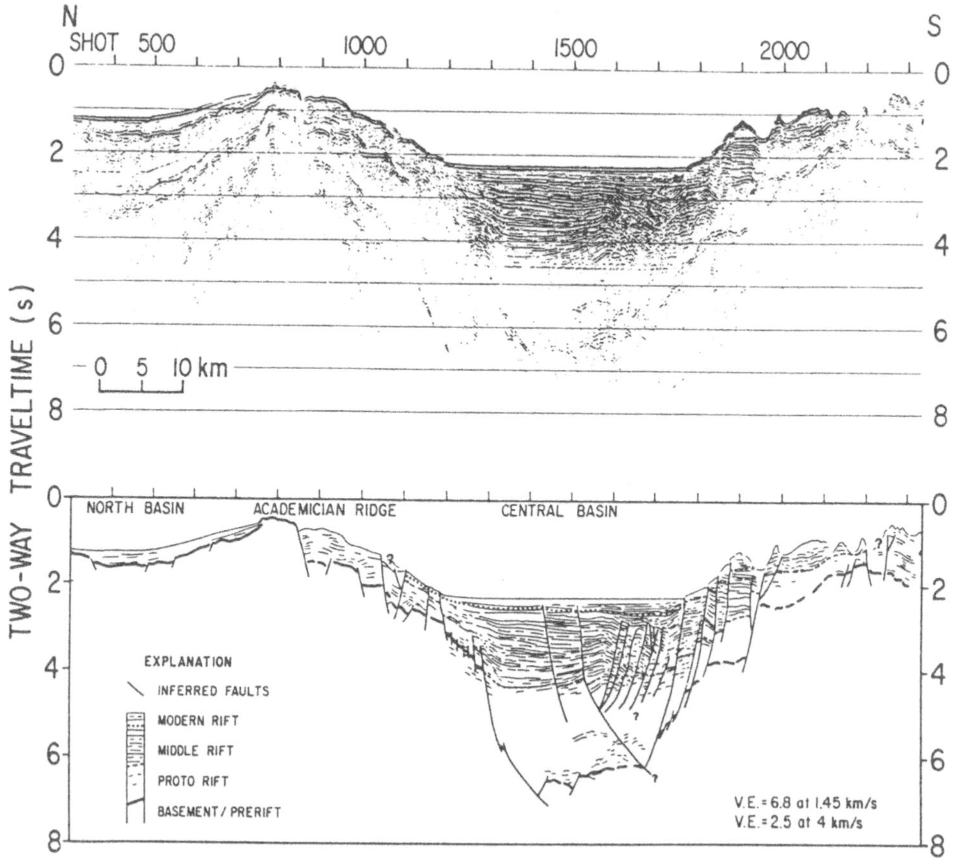

Fig. 2.
Multichannel seismic reflection line across the central part of Lake Baikal showing seismic data (top) and interpretation (bottom).
The thickest deposits are confined to a narrow trough, 15—20 km (9—12 miles) wide.

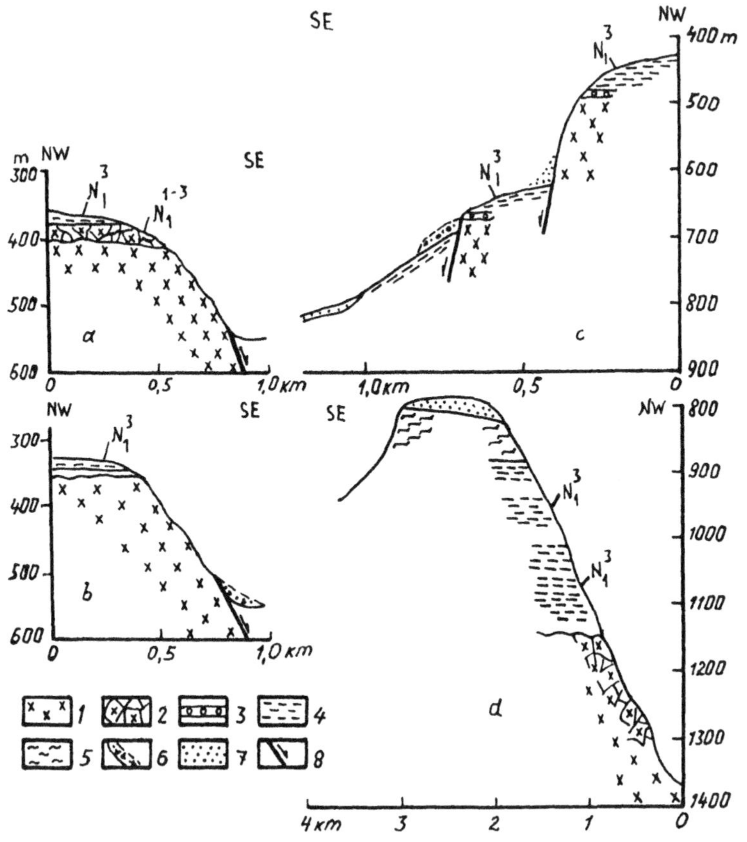

Fig. 3.
Geological cross-sections resulting from submersible studies at Academic Ridge:
a — across Academic fault, b — Primorsky fault, c — Olkhon fault,
d — underwater part of Svyatoy Nos Peninsula.
1 — metamorphic Precambrian rocks of the basement, 2 — wood debris,
3 — basal horizons, 4 — layered clays, 5 — non-layered clays,
6 — talus, 7 — Quaternary sediments, 8 — faults.

The first borehole 100 m long was drilled on Baikal in March of 1993. The specially designed lightweight drilling rig and hydraulic piston corer named "Baikal" were developed by "Nedra" Drilling Enterprise. The cores, 78 mm in diameter, were recovered (90%) and contained in plastic liners.

After the test drilling in Sljudinsky Lake in 1992, the first drill cores were obtained in the southern basin of Lake Baikal, in the site at the Buguldeika saddle, with 354 m water depth

(Fig. 4). The borehole was located in the vicinity of the Buguldeika River affecting the processes of sedimentation. The two stages (Manzurka, Neobaikalian) are related to the formation of the Baikal rift depression. The valley of Pra-Manzurka river, formerly the outlet of Baikal, formed at Manzurka stage, and Baikal water finally reached the Lena River. In the Neobaikalian stage (Late Pleistocene- Recent time, that is 300—500 Ma) the Buguldeika river developed its new thalweg towards Baikal. This stage can be divided into two phases. The first phase is known for downcutting of the river which was transporting coarser material, and the second one is characterized by a quieter stream and transportation of fine-grained material into Baikal.

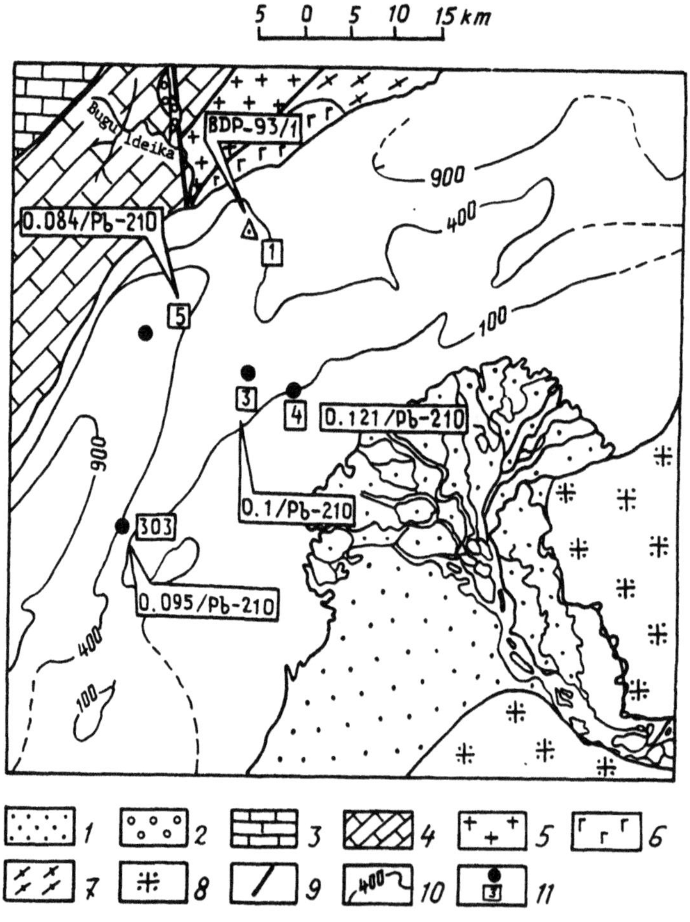

Fig. 4.
Schematic geological map of the Buguldeika borehole (BDP-93).
1 — Cenozoic deposits of Selenga River delta; 2 — Jurassic conglomerates, sandstones, coals; 3 — Upper Cambrian limestones, 4 — salt-bearing and carbonate deposits of the Lower-Upper Cambrian; 5 — Riphean granite-gneisses; 6 — ultrabasic and basic Riphean rocks; 7 — Upper Archean-Paleozoic gneisses; 8 — non-dismembered granitoids; 9 — major faults; 10 — isobaths, 11 — sites of piston core sampling, sedimentation rate values.

The high-resolution seismic reflection lines, obtained during the American-Russian expedition in 1992 in the region of Buguldeika, showed that the drill site was located on top of a tilted fault block within the Baikal Rift. The upper sedimentary sequence of 100 m meters thick is characterized by continuous subparallel reflections suggesting that sediments formed in lacustrine hemipelagic conditions. Some changes in the pattern of sediments are observed at 50 m depth, where variation of reflections characteristics takes place in the core.

Figure 5 presents the borehole section based on the primary core description. The sediments of this core primarily have silt-pelitic composition including lenses and interbeds of sand and silt, their amount increasing downcore. The cross-section represents a sequence accumulated by a continuous sedimentation from the water column. However, the patterns of sedimentation and composition of sedimentary material clearly show two sequences. The first, which is more fine-grained, displays a pronounced rhythmic structure. It apparently formed under relatively quiet conditions of lake sedimentation. The second, lower sequence is more coarse-grained with abundant turbidites. It includes more interbeds of sand and silt. This sequence is possibly related to the stage of an intense formation of the Baikal water discharge through Buguldeika River. This specific feature of the section of the BDP-93 borehole is confirmed by other investigations.

Geochemical data indicate that the entire sedimentary sequence penetrated by borehole is clearly divided into two parts: upper (over 49 m) and lower (below 49 m). The values of average abundance of elements, their dispersions and values of the Student t-criterion demonstrate significant differences between these two units. It is apparent that geological setting changed during the time of formation of the upper and lower sequences.

The investigations on distribution of organic compounds and diatoms as well as the palynologic (spore-pollen) analysis, are very significant to evaluate the climatic changes.

The downcore distribution of diatoms is shown in Fig. 5. It is evident that the diatom-bearing zones are alternated with diatom-devoid zones. These data on diatom distribution in the core correlate well with those on the distribution of biogenic silica. The peaks of biogenic silica and diatoms mostly occur in the upper 50 m of the core, whereas they are infrequent in the lower unit. A new diatom complex, never described in Baikal sediments previously, was found at 67 m depth. It consists of several species of *Stephanodiscus* genus. It was assumed that the age of sediments at this depth is not over 100000 years, since sediments of such age were collected by 10 m piston cores, but they do not contain diatoms of this species.

The palynologic analysis provides different types of spore-pollen spectra, e.g. "forest", "forest-steep" and "steppe" spectra. Based on this evidence, thirteen palynologic zones comparable with the periods of climatic variations have been recognized. These zones display alteration of moderate warm, relatively dry and moderately cold humid epoch in time of formation of sediments of the core.

Detailed paleoclimatic reconstructions need the determination of the absolute age of sediments. Unfortunately radiocarbon dates may be obtained only for the upper 5—10 m layer of sediments. Of 24 samples under treatment, the results are available only for 4 of them. They allow us to evaluate the sedimentation rate in about 17 cm per 1000 years for the last 15000 years. High sedimentation rates are substantiated by paleomagnetic data. The sediments of BDP-93 core correspond to Bruhnes, as their age is not over 500000 years. The further age determination will allow the detailed paleoclimatic analysis to be made from materials of the core section.

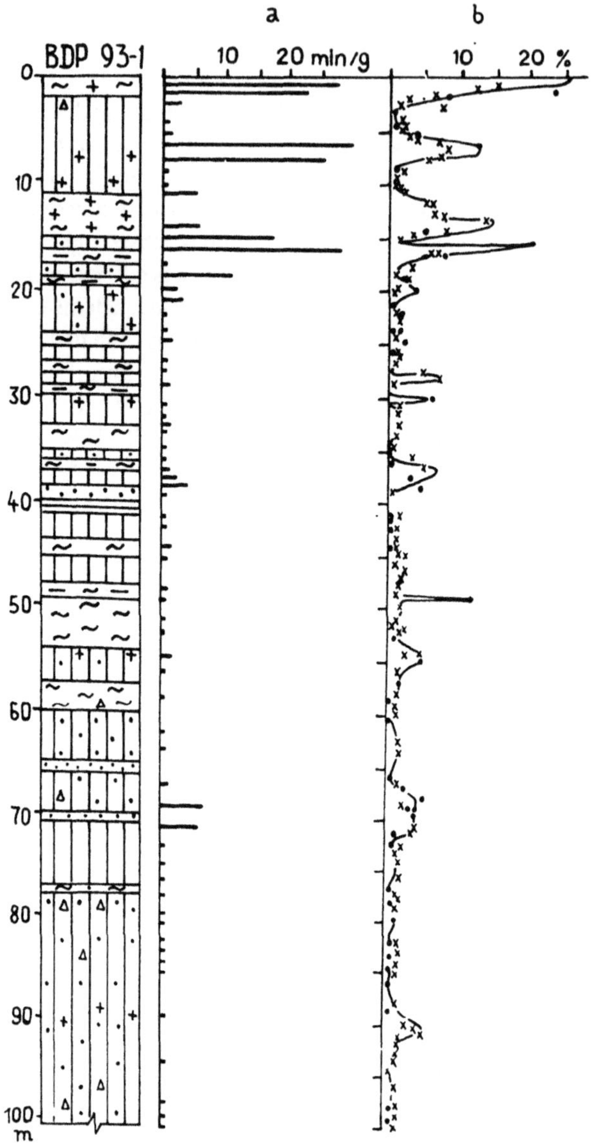

Fig. 5.
Lithologic section of the Buguldeika borehole (BDP-93) (right)
a — number of diatoms of the hole; b — biogenic silica content.
Short wavy line — pelitic silt, crosses — diatom ooze,
vertical hatching — clay, dots — sand, triangles — gravel.

It may be inferred that the material obtained from the investigations may be applied to the evaluation of the geological environment and climate changes. For this purpose large stratigraphic units may be used, e.g. the upper and lower parts of the drill core. To estimate climate changes it is necessary to study fluctuations in the content of organic material, rare elements, etc.

It is hopeful that in the future the project will be most effectively realized within the framework of International Programme "Mega Science Forum". This programme should be done in a step-wise manner: (i) drilling a number of 100 and 300 m boreholes at various water depth; (ii) superdeep drilling in the Tunka depression (over 7 km deep) and (iii) deep drilling in the Central basin of Baikal.

The Tunka valley which was formerly a lake, is situated west of Baikal, in the same Baikal rift zone. In Tunka, in contrast to Baikal, lacustrine sediments are alternated with volcanic rocks. The last volcanic eruptions occurred there only 50000 years ago. Thus, the study of the Baikal and Tunka depressions' evolution will help to evaluate that of amagmatic rift depressions with a wide development of volcanism.

The drilling and related works will permit us to solve a number of geological problems which are posed in the framework of continental drilling program as well as "Mega Science Forum", the latter discussed at the conference "Continental Scientific Drilling" in Potsdam at the end of August of 1993 [4].

The next main problems of modern geochemistry could be solved when the Baikal Drilling Project will be completely fulfilled:

— to establish the history of development of global changes of environment and climate in Inner Asia for the last 20—30 mln years;
— to evaluate the evolution of sedimentary basins related to rift systems;
— to create geological, geophysical and geochemical models of the Cenozoic cross-section of Central Asia exemplified by the Baikal region;
— to evaluate development of volcanism in the rift zone and to define thermal regime of the Interior;
— to study fluid regime of rift zones;
— to develop a new technology of drilling, in particular drilling from ice.

Finally, on the territory adjacent to the Tunka borehole, it is expedient to set up seismic polygon. The Stationary geological laboratory based at superdeep borehole together with the Siberian radiotelescope of the Institute of Solar-Terrestrial Physics, may serve as the grounds for complex study of natural processes at interrelated levels, i.e. space-atmosphere-Earth-superdeep hole.

It should be added that the success of this program lies in the fact that the Baikal rift zone is unique, and that there is already an excellent international team that has been working there fruitfully for some years.

References

1. Zonenshain L.P., Golmshtok A.Ya., Hutchinson D. (1992). The Baikal Rift Structure. Geotectonics, №5, pp. 63—77.

2. Hutchinson D.P., Golmshtok A.Ya., Zonenshain L.P., Moore T.K., Sholts K.A., Klitgord K.D. (1993). Preliminary Results of Multichannel Seismic Reflection Survey Carried out in Baikal Lake in 1989. Geology and Geophysics, №№10—11, pp. 25—36 (in Russian).
3. Kuzmin M.I., Williams D.F., Logachev N.A., Colman S. *et al.* (1993). The Baikal Drilling Program: Tasks, Proceeding, Solving Process Problems. Geologia i Geofizika, №№10—11, 5—15 (in Russian).
4. Deep Drilling. (1993). Paris, Organization for economic cooperation and development. p. 97.

♦

THE CASIMIR PROJECT: A JOINT RESEARCH PROJECT ON RIFT BASIN GEOLOGY BETWEEN SIBERIA AND BELGIUM

J. KLERKX
Royal Museum of Central Africa, Tervuren, Belgium

N.A. LOGACHEV and K.G. LEVI
International Centre for Active Tectonics and Natural Disasters,
Institute of the Earth's Crust of SB RAS, Irkutsk, Russia

V.D. ERMIKOV
Presidium of SB RAS, Novosibirsk, Russia

ABSTRACT. The long-term project CASIMIR was started in 1990 by joint efforts of geological institutes of SB RAS and Belgium and was devoted to the comparative study of sedimentary infill and mechanisms of rift basin formation, with Lake Baikal and East African lakes as examples. The number of participants in the project increased due to the involvement of scientists and students from European universities and research institutes of the RAS. Interdisciplinary research includes methods of structural geology, geophysics, geochemistry, geothermalism, limnology. The results obtained help to understand the processes of geodynamic evolution of the Earth, of the environment, global climatic change, origin of natural hazards, and other problems of social and economic significance.

Keywords: rift basins / international cooperation / interdisciplinary studies / Baikal / East African lakes.

The project CASIMIR is the geological facet of an agreement for scientific cooperation between the Siberian Branch of the Russian Academy of Sciences and the National Scientific Institutes of Belgium, devoted to the comparative investigation of Lake Baikal and the East African Rift Lakes. It stands for "Comparative Analysis of Sedimentary Infill Mechanisms in Rifts". It was stated in the agreement that the Project CASIMIR is part of the activities of the Baikal International Center for Ecological Research (BICER). Its activities are also connected to the International Center for Active Tectonics and Natural Disasters (ICATND).

The Project concerns all geological aspects related to the mechanisms of sedimentary fill in rift basins, from the tectonic aspect that involves the formation and evolution of the rift basins, to paleoclimatological and present global change aspects that are recorded in the basin sediments. Although the agreement directly involves the research institutes of the Siberian Branch of the Russian Academy of Sciences and the national scientific institutes of Belgium, it was agreed that other partners are invited to participate in the Project, particularly from the

V A Koptyug and J Klerkx (eds), Science Policy New Mechanisms for Scientific Collaboration between East and West, 117–125

Russian side, scientists and organizations from Siberia and other parts of Russia, and from the Belgian side, scientists from Belgian universities as well as scientists from other countries.

The agreement for scientific cooperation was signed in 1989 and the Project started in 1990 with a preparatory stage that consisted of a reconnaissance of the scientific and logistic possibilities of both partners, expert visits and meetings, as well as joint field excursions around Lake Baikal, and in to the region of Lake Tanganyika and Lake Malawi in Africa, and identification of scientific items and preparation of programs. In 1991 the field and laboratory investigations started. Joint teams mainly consisted of scientists from the Institute of the Earth's Crust at Irkutsk (Russia), and from the Royal Museum of Central Africa at Tervuren (Belgium), and included from the beginning scientists from Russian organizations and from Belgian and European universities. From the Russian side scientists became involved from the Limnological Institute at Irkutsk, the United Institute of Geology, Mineralogy and Geophysics at Novosibirsk, as well as the Geological Institute at Moscow. From Belgium, scientists from the University of Ghent also participated in the scientific program.

The scientific orientations, defined in the initial stage of the Project, correspond to three main scientific topics (Fig. 1):

- the structural geology of the basement surrounding the rift basins in function of its control on rift basin evolution;
- the recent faulting controlling the rift basins, its present state and evolution during time;
- structural processes within the basins and mechanisms of sedimentation.

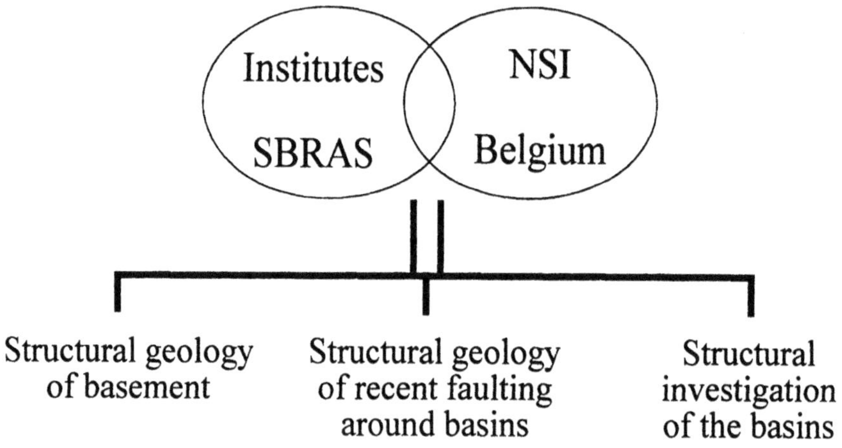

Fig. 1.
The thematic content of the CASIMIR project in the initial stage (1991).

The main characteristics of the initial stage of the Project CASIMIR were:

- that joint teams were operating since the very beginning, in Lake Baikal as in Africa;
- that a broad approach was considered, involving several aspects;
- that these aspects mostly belong to the structural features of rift basin geology, and furthermore concern a classical approach to rift problematics.

The cooperation became efficient soon after its initiation, and scientific results were already obtained after the first operation. This efficient start was mainly due to:

- the infrastructure available in the Siberian Institutes, particularly for the operations on Lake Baikal, and the experience of their staff;
- the geological similarities between the two investigated regions;
- and the successful way in which the teams complemented each other in knowledge, experience, and technique.

It should also be noted that, although in this stage the Project focused on the classical aspects within the problematics of rift basin evolution, different subdisciplines within the earth sciences were involved in the research program. This approach was highly favored due to the experience of scientific teams of the Siberian institutes for a group of different subdisciplines working within a single program towards a common objective. This interdisciplinary approach had been applied before for the investigations in the Baikal area. For the Belgian partners, due to the limited size of their scientific community, the cooperation with the Siberian institutes offered the possibility to introduce techniques and approaches that had not been available earlier.

In its later stage, the growth of the CASIMIR project was characterized mainly by:

- the introduction of scientists beyond the main teams that initiated the Project. Scientists from other institutes in Siberia and Russia joined the Project, as well as scientists from Belgian and European universities;
- an improved interdisciplinarity. This concerns first of all the connection with other subdisciplines in the earth sciences, and secondly interaction with other disciplines. This concerns particularly the investigation of the lake basins, where links were established by teams studying related problems within the physical and chemical limnology;
- another important achievement in this stage: whereas before it had been limited to a classical approach of the problem, now the project explored the socioeconomic aspects of the earth sciences, more particularly environmental problems and natural hazards. This was especially furthered by interaction with other disciplines related to the physical and chemical investigation of the lakes.

Main scientific programs and achievements

1. THE RELATIONSHIP BETWEEN RECENT RIFT STRUCTURES AND BASEMENT TECTONICS

One of the main characteristics of the Baikal and Tanganyika-Malawi rifts is that their structure parallels the basement structural pattern. A relationship between both has been claimed since long. The Belgian team has been investigating the mechanisms of this reactivation process several years in the African rift, and has been able to show that the leading structure of the present rift architecture in this region has been generated by a major zone of weakness of the Earth's crust, dating from two billion years ago, and that this zone of weakness has been repeatedly reactivated during geological times. In this zone, lubricated corridors have been created within the Earth's crust, which have been reused by the recent rift faulting. A very similar mechanism has been identified for the Baikal rift; the present rift architecture is also imposed on a major dislocation zone that originated about two billion years ago.

This reactivation mechanism is an important contribution towards understanding the origin and the evolution of deep rift basins, not only in the rifts themselves, but also in the problematics of active or passive rifting.

Active rifting in continental lithosphere is considered as a result of extensional geodynamics due to a local deep-seated source of energy in the form of astenosphere upwelling. Passive rifting may be defined as an extension of the lithosphere and rift basin development, caused by external forces that influence a larger area and produce continental rifting, a secondary tectonic phenomenon. This last solution has been suggested for the Baikal rift, considering that it originated from the transfer of stresses in the Asian continent as a result of the India and Eurasia collision along the Himalaya front.

2. Recent evolution of the rift basins

The deep rift basins are ancient structures which formed 30 million years ago. Their internal structure and evolution are still poorly understood. The Project focused on the distribution of the stress field within the rift structure, at the present time, and in the past. This study has shown that in both the Baikal and African rifts stresses vary, from place to place and have changed during the course of rifting. These results contribute definition of the forces that were responsible for basin formation and evolution in the past, to understand the present evolution of the basins, and even possibly to the prediction of their future evolution, leading to continental break up. Indeed, this study has shown that the rift basin evolution in Baikal and in the African rift is complex temporally and developed over a long time. The results also contribute for understanding the formation of old sedimentary basins of rift type, where thick accumulation of sediments favor hydrocarbon formation potential.

3. Heat flow and hydrothermal activity in rift basins

The systematic heat flow mapping of Lake Baikal over many years has shown nearly normal heat flow in the major part of the basin, as well as very strong heat flow anomalies locally, which appear to be linked to fault zones, and which recently have been correlated with hydrothermal discharge on the lake floor. Heat flow measurements in Lake Tanganyika and Lake Malawi, although preliminary, evidence a similar situation. Consequently, it appears that the contribution of convective processes in the heat balance in deep rift basins compared to conductive processes is more important than is generally accepted. In Lake Baikal, the bottom hydrothermal discharge was investigated by combining heat flow probing, detailed bottom bathymetry, and definition of the physical parameters of the water column. The discharging hydrothermal water was shown to remain as a dense layer near the lake bottom, and to become concentrated towards the deeper parts of the lake. This process contributes to the ventilation of the Baikal water mass.

4. NATURAL HAZARDS IN A RIFT ENVIRONMENT

In the region of the northern Malawi rift, particular attention has been paid to the natural hazards that are connected with active faulting. Different types of risks are concentrated in this region: the seismic risk was investigated by a local seismic network, which provides new information about the distribution of earthquakes and the geometry of active faults. Combined field work and analysis of satellite imagery highlighted the active faulting and basin tilting in the northern extremity of Lake Malawi. Furthermore, high resolution seismic profiling in the northern lake basin has evidenced extensive accumulation of gas in the shallow sediments of the basin, and the escape of gases along fault zones. The gases consist of a mixture of CO_2, and methane. The nature and the origin of hydrocarbon gases were investigated in the water column and in the sediments.

5. VOLCANISM AND DEEP PROCESSES UNDER THE RIFTS

Volcanism is relatively poorly represented in the Baikal and Tanganyika-Malawi rifts. Nevertheless, investigation of its spatial and temporal distribution, and geochemical composition is an important factor in understanding the processes that affect the Earth's crust and mantle beneath the rifts. A geochemical study of the volcanic fields around Lake Baikal is underway. Particular attention is being paid to the ultramafic nodules contained in the lavas, which record the mantle processes beneath the rift region.

6. PALEOHYDROLOGIC RECONSTRUCTION OF NORTHERN LAKE MALAWI: THE SULFUR AND CARBON CYCLE IN THE LAKE SYSTEM

The Lake Malawi water system is investigated to study the cycle of some chemical elements and their isotopes, in order to precisely locate the lake water balance and to establish exchange phenomena and physical properties of the water column. The study aims at a paleohydrological reconstruction of the lake and the paleoclimate evolution of the region.

Present state of the Project

At its present state the Project has evolved from a rather narrow and classical approach of the rift problematics, from the geological point of view (Fig. 1), towards a project where numerous and various aspects, within the geosciences framework are considered (Fig. 2):

- geophysics:
    - seismic profiling,
    - seismology,
    - heat flow;
- structural geology:
    - basement structure,
    - recent faulting,
    - stress field evolution;
- geochemistry:

- nature and origin of the volcanics,
- mantle xenoliths;
- hydrothermalism:
  - hydrothermal outflow,
  - gases.

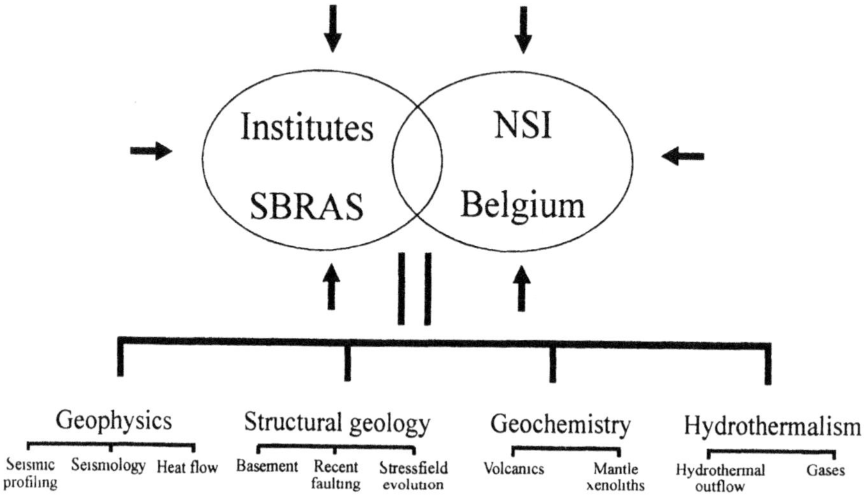

Fig. 2.
The scientific content of the CASIMIR project at a later stage (1993-1994).

All these aspects employ different techniques and scientific approaches towards the same objective: a better understanding of the processes of faulting, currently active in the rift basins, which define the origin and shape of the basins, and are responsible for their evolution. These approaches also attempt a better understanding of the processes which affect the Earth's crust and mantle in a rift region. Consequently, an interdisciplinary approach between different subdisciplines within the field of geosciences is necessary for the solution of the same geological goal.

This rapid development of the Project, due to its multidisciplinary character, was aided by the opportunity to lean on the research institutes of the Siberian Branch of the Russian Academy of Sciences, which is based on a multidisciplinary structure. Therefore the rapid involvement of different disciplines was a normal approach. Furthermore, through the combination of different teams the Project profited from an addition of knowledge, experience and techniques.

Another unique characteristic of the Project is that it is based on the involvement of permanent teams, the research institutes of the Siberian Branch on the one hand, and a permanent Research Institute in Belgium on the other hand. This allows the existence of permanent teams, and permits the construction of long-term projects, which rely on a continuity in personnel and structure. However, it is evident that these teams are not able to tackle all necessary aspects, and therefore it is important that external teams and individuals join the Project with their particular techniques and approaches, for a long term involvement or for selective intervention in the function of a particular problem.

At present the following institutions are participating in the CASIMIR project:

From the Russian side:

- the Institute of the Earth's Crust at Irkutsk and the connected International Center for Active Tectonics and Natural Disasters,
- the Limnological Institute at Irkutsk and Baikal International Center for Ecological Research,
- the United Institute of Geology, Mineralogy and Geophysics at Novosibirsk.

Are also participating:

- the Geological Institute at Moscow,
- the Institute of Oceanology at Moscow,
- the Institute for Water and Environmental Problems at Barnaul.

From the Belgian side:

- the Royal Museum of Central Africa at Tervuren, with particular interventions from:
- the Free University of Brussels,
- the University of Ghent.

Different European groups have joined for specific topics:

- the Free University Amsterdam (The Netherlands),
- the University of Paris-Sud (France): Laboratory of Hydrology and Isotope Geochemistry,
- la Laboratoire de Géologie du Quaternaire at Marseille (France),
- the Federal Institute for Water Resources and Water Pollution Control at Zurich (Switzerland).

Besides the multidisciplinary approach within the geosciences, another important achievement of the Project is its exploration of adjacent disciplines. The main target of the geoscientist is to investigate in rift problematics those aspects that are related to the lake basins and their evolution, but he cannot escape the fact that these basins are water reservoirs, consequently the geological aspect encounters problems that are related to physical and chemical limnology (Fig. 3). The Project used the opportunity to connect with these disciplines, and programs are developing where interactive, multidisciplinary investigations are promoted.

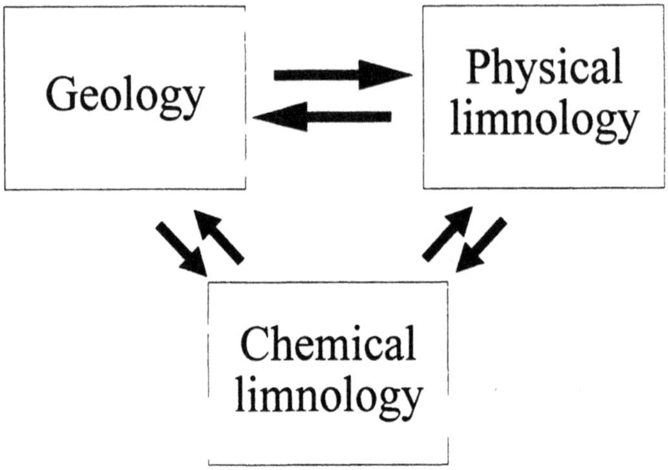

Fig. 3.
Interaction of disciplines within the CASIMIR project at a later stage.

Future development and evolution of the Project

After a few years of scientific activity the CASIMIR project requires an evaluation. This evaluation has to take into account the quality of the different activities, their efficiency, but also their significance and their importance in relation to a goal-oriented approach. In fact, the project started without particular priorities, without specific objectives, without well-defined time limits. All the programs were initiated within the general objective of rift basin geodynamic evolution. This approach appeared to be efficient, to allow different programs to start, and to reach achievements in a short span of time. Nevertheless, at this stage, priorities need to be defined and programs need to be elaborated with the help of evaluated priorities. Profiting from its structure, especially the fact that the Project can rely on permanent research teams, it needs to combine the advantages of a long-term project and those of a problem solving, target-oriented, approach. Without neglecting the fundamental scientific aspects related to rift basin development, the Project has also to consider those aspects within the geosciences that are of socio-economic significance (Fig. 4).

These aspects are found at the interface with other disciplines, particularly physical and chemical limnology. Developing scientific activities in combination with other disciplines will lead to new developments by combining the techniques and the scientific approach of two or more disciplines. It requires creativity and innovation, but the Project has already decided upon this course. For example, the investigation of the hydrothermal activity in the Baikal Lake basin has opened new developments for water physics and chemistry in close connection with limnologists; combining approaches and techniques from different disciplines appears to be highly beneficial. In the Lake Malawi basin, the fundamental results for basin geometry and structural evolution, translated in environmental terms, benefit the hydrological investigations and paleoclimatic reconstructions. Finally the main aim and basic content of the CASIMIR project should be defined as an interdisciplinary evaluation of environmental changes and natural hazards in active zones of Cenozoic continental rifting.

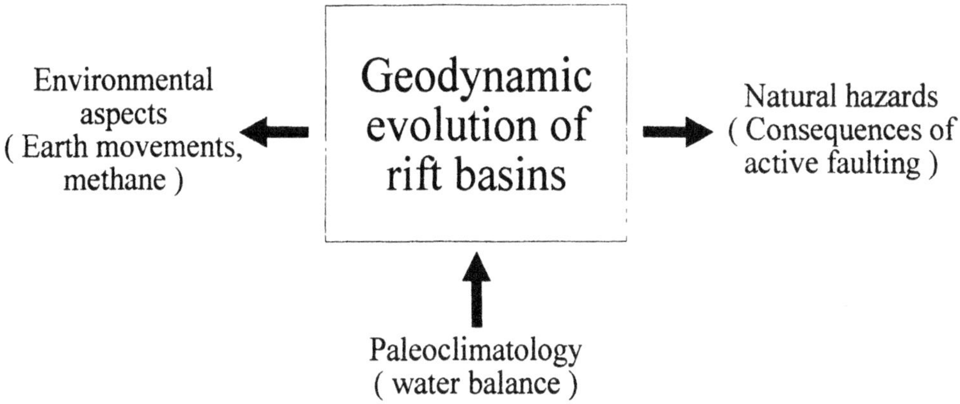

Fig. 4.
Socio-economic facets of the CASIMIR project.

The problematics of the CASIMIR project belong to the fundamental sciences, but its interdisciplinary and multidisciplinary approach extends to the socio-economic aspects connected with the large lake basins. Therefore the Project is able to combine fundamental research related to the particular problem of large and deep continental rifts, and goal-oriented research answering fundamental questions about the future of our human society, and environmental and global change.

II.3. NATURAL SCIENCES

AND TECHNOLOGICAL APPLICATIONS

INTERNATIONAL COOPERATION IN THE FRAMEWORK OF THE SIBERIAN SYNCHROTRON RADIATION CENTRE

G.N. KULIPANOV
Siberian Synchrotron Radiation Centre,
Budker Institute of Nuclear Physics of SB RAS, Novosibirsk, Russia

ABSTRACT. In the last decade the Siberian SR Centre has implemented a wide program of research in cooperation with various research centres and institutions in UK, Germany, France, India, Korea, Belgium, and the USA. The report illustrates the diverse kinds of cooperation, including joint experiments with the use of SR sources available at the Budker Institute of Nuclear Physics, the implementation of new joint projects in Novosibirsk and in the other centres, as well as the delivery of equipment to foreign countries, designed and manufactured at the Budker INP or in collaboration with Russian industry.

Keywords: synchrotron radiation centre / free electron laser / wigglers / undulators / synchrotron radiation experiment.

1. At present the Siberian Synchrotron Radiation Centre (SSRC), which was organized on the basis of the Budker INP laboratories, is a major site for synchrotron radiation and free electron laser (FEL) researches in Russia. The general lines of the SSRC scientific program are:
- the performance of experiments and the development of new technologies using synchrotron radiation from the Budker INP SR sources - the VEPP-2M, VEPP-3 and VEPP-4 storage rings;
- the creation of experimental equipment for SR research (beamlines, optics, onochromators, detectors);
- the development of storage rings - dedicated SR sources, the creation of devices for SR generation - wigglers and undulators;
- the creation of free electron lasers.

SSRC has no budget source of funding and performs its experimental and research programmes at the expense of:
- the State scientific-technical program "Synchrotron radiation";
- purposeful funding of FEL works;
- grants of Russian scientific and technological funds;
- contracts with various institutions in Russia and abroad;
- international cooperation.

2. The SSRC activities cover a wide spectrum of scientific and technological problems (Fig. 1). The investigations in the VUV and soft X-ray ranges ($\lambda = 2.10^3 \div 3$Å) are being

V. A. Koptyug and J. Klerkx (eds.), Science Policy: New Mechanisms for Scientific Collaboration between East and West, 129–140.

carried out at the VEPP-2M storage ring (0.7 GeV energy). For works in the X-ray range ($\lambda = 3 \div 0.3$Å), the synchrotron radiation generated by VEPP-3 (2 GeV energy) is utilized. In 1983—85, the VEPP-4 storage ring (6 GeV energy) was used in hard X-ray experiments. These works are expected to be resumed in the fall of 1994 in a new experimental room, which is being constructed. Nineteen experimental stations, which now operate at the VEPP-2M and VEPP-3, are listed in Table 1. SSRC works were presented in over 1100 publications, in Russian and foreign journals [1]. Fig. 2 gives information concerning the annual number of SSRC publications. Needless to say, the subgects of these papers are noted for great diversity.

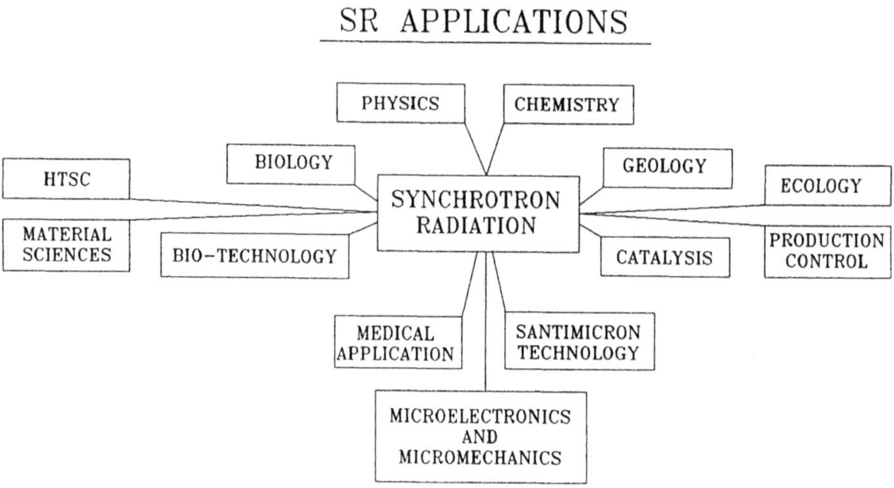

Fig. 1.
SR applications in SSRC.

X-ray lithography	Digital subtraction angiography
Photoelectron spectroscopy	X-ray microscopy and microtomography
Photodesorption	Time resolved powder diffractometry
Time resolved luminescence	Macromolecular crystallography
VUV and SXR metrology	Time resolved X-ray luminescence
Optical klystron	Small angle scattering
Compton back scattering	Topography and diffractometry
Laue diffractometry	EXAFS spectroscopy
Anomalous scattering	LIGA technology
X-ray fluorescence elemental analysis	

Table 1.
The list of SSRC experimental stations.

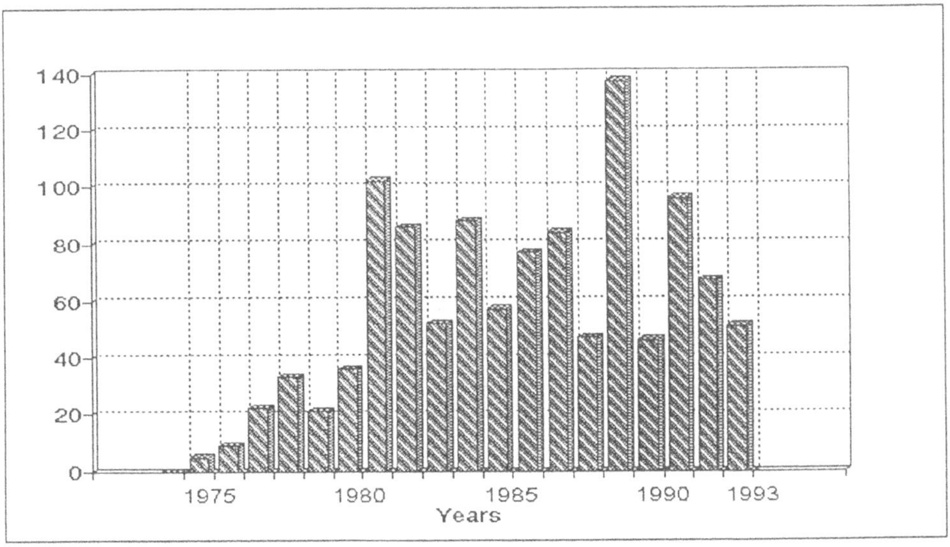

Fig. 2.
The annual number of SSRC publications.

In works that apply X-ray diffraction methods, the structural changes in metals in the process of their destruction, in solids during chemical reactions and in the contracted muscle, as well as the phase transitions under superhigh pressures are studied. Due to EXAFS spectroscopy, the structure of amorphous semiconductors and metal glasses, the active centres in proteins and various catalysts are examined. Among applications of X-ray fluorescence elemental analysis are a search for new ore deposits, analysis of marine water and sediments, elemental analysis of aerosols, microelemental analysis of water, medicines, blood, and so on. Synchrotron radiation allows one to examine the structure of the surface layers of materials of submicron thickness, to take two-microns-resolved microtomograms (for example, of lymphonodes), to measure the concentrations of elements in fluid inclusions in minerals, to measure the distribution of various elements along the humain hair, and so forth. Using the X-ray lithographic technology, regular micropore filters with high — up to 50% — transparency, with given arrangement and sizes of 4 - mm - diam pores on a 26 - mm - thick lavsan film were manufactured. And the X-ray lithographic technology for manufacturing field transistors whose channel is 0.25 mm long was improved.

The staff of the experimental teams involved in SSRC works is mainly from Novosibirsk institutes of the Siberian Branch of the Russian Academy of Sciences. These teams are usually the hosts of the experimental stations. Besides, the teams from the institutes of the

Russian Academy of Sciences in the other cities, universities and higher education schools, technological institutes from industry both in Russia and abroad. Fig. 3 illustrates the diagram which shows the dynamics of changing the number of the experimental teams since 1973, for two decades. The decreased number of teams in 1986—87 was caused by the fire at the institute in August of 1985. The second reduction after 1990 was due to the change in the political and economic situation in the countries of Eastern Europe and in the republics of the former Soviet Union. As a consequence, 10 teams from the former German Democratic Republic, 6 teams from the former Czechoslovakia, 3 groups from Hungary, as well as the teams from Armenia, Azerbaigan, Estonia, Latvia and Ukraine stopped working. A drastic reduction of funds for science in Russia and the increased prices of flight tickets and hotel accommodation brought about the decrease in numbers of teams from cities in the European part of Russia. In the current year, as few as 64 teams have worked at the SSRC, 12 from foreign countries (2 from UK, 3 from Germany, 3 from Korea, 1 from India). The list of agreements on scientific cooperation between SSRC and foreign laboratories is given in Table 2, with the field of cooperation indicated.

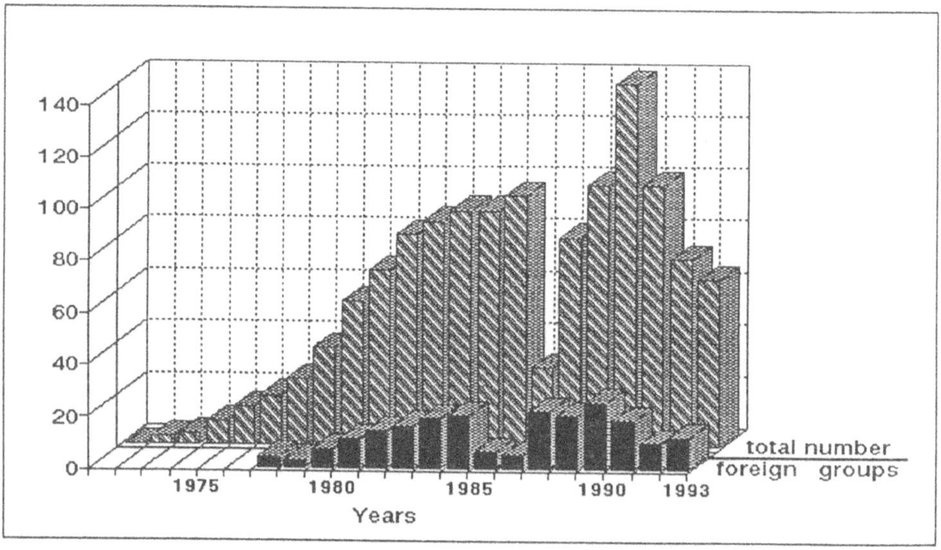

Fig. 3.
Variations in the number of users teams in SSRC.

1. Daresbury Lab (SERC United Kingdom)	1977	Production and utilization of SR
2. LBL, SLAC University of Wisconsin (DoE USA)	1980	SR beam production insertion devices, compact sources, development of SR experimental equipment
3. CAT, BARC, Inter-University Consortium (DST India)	1987	SRS design and development, SRS utilization
4. FEL Lab Duke University (NC USA)	1992	Free electron lasers
5. POSTECH (Korea)	1992	Accelerator design, beam lines insertion devices, SR experiments
6. IHEP Academia Sinica (China)	1992	Scientific exchange and Joint Research and development in the field of SR
7. Laboratory BESSY (Germany)	1993	Development, design and construction of BESSY-II
8. APS ANL (USA)	1993	Area synchrotron radiation instrumentation

Table 2.
The list of collaboration agreements between SSRC and foreign laboratories.

3. In the last decade, SSRC has executed contracts on the design and manufacture of dedicated SR sources [2] — the electron storage rings SIBERIA-1, TNK (ZELENOGRAD) and SIBERIA-2 (their parameters are listed in Table 3). Since 1983 the SIBERIA-1 storage ring operates at the Kurchatov Institute (Moscow). As for SIBERIA-2, this storage ring is fully assembled at the Kurchatov Institute. The difficulties with funding for the last years has somewhat delayed its assemble and commissioning. However, we hope to have the electron beam in SIBERIA-2 by the end of 1994. The TNK storage ring, intended for the Technological Microelectronics Centre in Zelenograd, is also ready for assemblage, but all its components are still stored at Novosibirsk. The work on the creation of this centre was ceased because of the lack of funds.

The same reason also stopped the SSRC programmes on the creation of compact SR sources on the basis of superconducting magnets [3]. This programmme has envisaged the modulus principles when designing technological storage rings whose key component is magnets of one and the same type, namely short rectangular magnets with a 6 T homogeneous magnetic field. The layout of these magnets makes it possible to construct, from 8 magnets, a compact 600 MeV storage ring for submicron lithography ($\lambda \sim 8Å$), composed of either 16 magnets — the 1.2 GeV storage ring for LIGA-technology to manufacture micromechanics elements ($\lambda \sim 2Å$), or 24 ones — the 2.4 GeV storage ring for angiography ($\lambda \sim 0.5Å$). A mixed magnetic structure, which comprises conventional 1.6 T and superconducting 6 T magnets positioned in minimum betatron functions, open up new possibilities. Such a structure ensures low emittance of the beam, high brightness and a wide spectral range suitable for various kinds of experiments.

Name	"Siberia-1"	"Zelenograd"	"Siberia-2"
Status	Operates	—-	Constr.-1994
Energy (GeV)	0.45	1.5	2.5
Emittance ($nm \cdot rad$)	880	27	78
Critical wavelength (Å)	61-BM,21-W	8-BM	1.75-BM
Stored current (mA)			
total	360	300	300
single bunch	360	100	100
SR power (kW)	4	$\sim$ 30	200
RF frequency (MHz)	34.5	181.3	181.4
Bunch length $2 \cdot \sigma_s$ (cm)	60	4	4
Lifetime (h)			
touchek	5	10	10
vacuum			
Type of filling	e^-	e^-	e^-
Number of beamlines	7	39	39
Insertion devices			
number	1 (W)	2 (SCW) 5 (W) 2 (U)	7 (2W+5U) 2 (SCW)

Table 3.
The parameters of the dedicated SR sources:
SIBERIA-1, TNK (ZELENOGRAD) and SIBERIA-2.

The SSRC activities on the creation of dedicated SR sources has covered (Tables 4 and 5):

— the development of the conceptual project of a SR source for LIGA-technology, for KfK (Karlsruhe, Germany);
— the design of the components of the magnetic system for the synchrotron-injector and the main storage ring BESSY-2 (Berlin, Germany);
— the participation in the projects for new SR sources SLC (PSI, Switzerland) and DIAMOND (Daresbury, England), which underlie the Novosibirsk idea of a mixed magnetic structure and the Novosibirsk design of a superconducting magnet [3].

1. Optical mirrors for UVSOR	June 1991 - Oct 1991	Institute for Mol. Science (Japan)
2. Design, construction and comissioning RF cavity for Duke storage ring	Nov 1991 - Dec 1993	Duke University (USA)
3. Conceptional design of SR source for KfK	March 1992 - May 1992	KfK Karlsruhe (Germany)
4. 7.5 Tesla wiggler for Pohang light source	Aug 1992 - Oct 1994	POSTECH, POHANG (Korea)
5. Conceptional design FEL for project SELENA	June 1993 - Dec 1993	NASA Duke University (USA)
6. Design of magnetic components of th BESSY-2	Nov 1993 - Feb 1994	BESSY Berlin (Germany)
7. Manufacturing of prototype elements beam lines APS	Sept 1993 - Dec 1994	APS ANL (USA)
8. Design and manufacturing wiggler for production of elliptically polarized SR	Dec 1993 - June 1994	APS ANL (USA)

Table 4.
The list of contracts between SSRC and foreign customers.

1. Research and Development of superconducting magnets as sources of SR in electron storage rings	Dr V.Suller Daresbury Lab (United Kingdom), Prof E.Jaeschke BESSY (Germany), Dr Mezentsev INP (Novosibirsk)
2. Developments on the existing storage ring FEL in UV and prospects concerning advanced facilities	Dr M.E.Couprien CEA-LURE (France), Dr G.Luycky NIKNEF (The Netherlands), Dr N.Vinokurov INP (Novosibirsk)
3. Development of basic studies in the field of micro- and nano-fabrication using e-beam lithography and X-ray lithography	Dr H.Launois CNRS LZM (France), Dr G.Kulipanov INP (Novosibirsk), Dr A.Aseev ISP (Novosibirsk)
4. Multilayers for intense X-ray beams: new materials, study physical processes in films and interfaces	Dr P.Dhez Univ.Paris Sud (France), Dr J.Walker ELLETRA (Italy), Dr N.Chkhalo INP (Novosibirsk)
5. Atmospheric aerosols in the Asian FSU	Prof R.Van Grieken Univ.Antwerp (Belgium), Prof R.Jaenicke Univ.Mainz (Germany), Prof K.Koutsenogy ICKC (Novosibirsk)
6. The fast two-coordinate X-ray detector with advanced frame-by-frame regime	Dr R.Gehrke HASYLAB, Prof J.R.Helliwell Manchester University (United Kingdom), Dr. B.Tolochko ISSC (Novosibirsk)

Table 5.
The list of proposals to the International Association for the promotion of cooperation with SSRC.

4. The INP has designed and manufactured a great deal of dedicated SR generators — wigglers and undulators — on the basis of superconducting magnets, permanent magnets and conventional electromagnets [2]. The first in the world superconducting 20-pole wiggler, installed at the VEPP-3 storage ring in 1979, increased to increase the brightness of the source within λ ~ 1Å by a factor of 200. The superconducting wiggler with a record field of 8 T, mounted at the VEPP-2M storage ring in 1982, made it possible to generate, at a 0.7 electron energy, intense synchrotron radiation in the 35Å internal. In 1983, the helical undulator was located at the VEPP-2M storage ring, which was, for a long period of time, the only source in the world for quasimonochromatic undulator radiation possessing circular polarization. Moreover, the INP scientists were the first who invented, designed and built undulators on the basis of Sm-Co permanent magnets with the use of iron poles for the concentration of the magnetic flux in the longitudinal direction. They were mounted at VEPP-3 in 1979—1981. Known as "hybrid", undulators of this type are in intense use at many SR centres all around the world.

In recent years the SSRC is a participant in the development and manufacture of wigglers for the centres abroad:

— a superconducting 7.5 T field wiggler for PLS (University in Pohang, Korea);

— a prototype of the wiggler for the production of circular polarized radiation on APS (ANL, USA), which will be tested at the NSLS (BNL, USA) in 1994.

5. FEL investigations are being carried out in two directions:

— FEL experiments in the visible and UV ranges on the VEPP-3 bypass;

— the creation of a powerful FEL in the IR region (320 mm) at an average power of up to 100 kW on the basis of a new accelerator — microtron-recuperator.

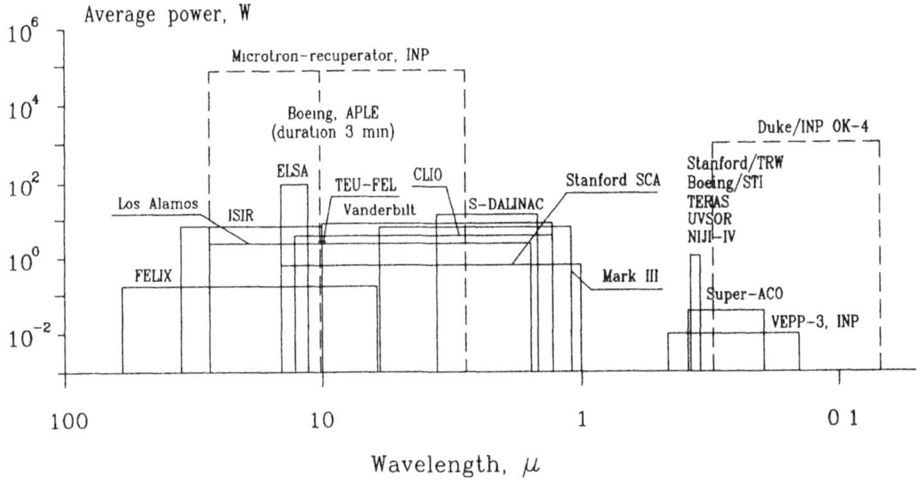

Fig. 4. Status of creating FEL in the world.

In 1989, a FEL — the optical klystron of the visible and UV range (0.7 ÷ 0.24 mm) was constructed at the VEPP-3 bypass. So far this is the only FEL in the world, which operates in the UV spectral range (Fig. 4). On this device, experiments with the intracavity Fabri-Perrot standard have recently been performed and the generation whose monochromaticity is 210 has been obtained. The series of experiments on the generation of the optical klystron in the confocal optical cavity is completed. This is of significance for the creation of long optical cavities of powerful FEL. In addition, the experiments have been conducted concerning the interference of radiation from two undulators separated by the achromatic bend with a view to simulate the "electron radiation extraction" from a high power FEL.

The works on the VEPP-3 bypass are expected to be accomplished in 1994. Jointly with Duke University (USA), we prepared the project "Short wavelength FEL development and application project". The project envisages FEL transportation from the VEPP-3 bypass to Duke University, its installation at the dedicated Duke electron storage ring, as well as obtaining new physical results on generation within the $0.24 \div 0.05$ micron region (Fig. 4). The American side funds this project within the frames of "Medical FEL program", and also, possibly, within the frames of the International Centre of Science and Technology.

The second direction of FEL activities deals with the creation of the Centre of Photochemical Researches in Novosibirsk. In the end of 1992, the Decree of the Presidium of the Siberian Branch of the Russian Academy of Sciences on the creation of this Centre on the basis of the Institute of Chemical Kinetics and Combustion and the Budker INP was signed. For the Centre, one of the buildings of the Institute of Chemical Kinetics was allotted, whereas 350 micron FEL at an average power of up to 100 kW will be housed. This device has been designed and built by the INP for the last few years. Now the Institute of Chemical Kinetics and Combustion together with some other institutes of the Siberian Branch of the Russian Academy of Sciences prepares the groundwork for the employment of this laser for solving the problems in photochemistry. The FEL is planned to be commissioned in 1996.

The project is funded at the expense of the Russian programme on FEL creation and also at the expense of the contracts with foreign institutions (Table 4). In addition, the international cooperation program is being prepared.

In July of 1993, the Russian-US Workshop on power beaming of satellites, including geostationary ones, for the satellite information network was conducted. The meeting was attended by the representatives of the INP, the Institute of Chemical Kinetics and Combustion (Novosibirsk), Experimental Design Offices "Raduga", "Granat" and "Astrofizika" from the Russian side and those of NASA, Navy laboratory in China Lake, Duke University, Los Alamos laboratory and Boeing Company from the American side. The workshop made a decision to prepare the joint project "FEL power beaming" on the basis of the FEL which is being designed at the Budker INP [4] and the USA programme SELENA [5].

Besides, for the ARPA agency the joint Russian-US proposal "Applications of Defense Laser Technology to the remediation of toxic and hazardous wastes" was prepared. The proposal was prepared by the Institute of Chemical Kinetics, Duke University, the State North Carolina University and some other USA institutions.

The major result of the Russian-US cooperation in the field of FEL might be, in the not too distant future, the completion of the creation of the FEL at an average power of 100 kW and its testing in Novosibirsk, with the participation of the Russian and American partners, in order to check the main principles underlying the project and to demonstrate the serviceability of the FEL.

The more distant result of this cooperation might be the creation, by Russian and American partners, of FELs at higher average power ($200 \div 1000$ kW).

In addition to the VEPP-3 bypass and FEL works for the Photochemistry Centre, the INP considers the project of a compact FEL at an average power of 1 W within 1 ÷ 50 mm. This project is suggested to implement jointly with CAT (Indor, India). Also considered is the project of a free electron laser of visible and UV ranges at an average power of 1 kW. This FEL could constructed on the basis of a dedicated storage ring having strong-field superconducting magnets. This project is assumed to be implemented in the framework of the cooperation between Russia on the one side and France (CEA-LURE) and the Netherlands (NIKNEF) (Table 5).

6. The Siberian Synchrotron Radiation Centre conducted regularly All-Union conferences on Utilization of synchrotron radiation (SR-75, 77, 78, 80, 82, 84, 86, 88 in Novosibirsk, and SR-90 in Moscow). The proceedings of the SR-82, 84, 86, 88, 90 conferences were published either as collected books [6, 7, 8], or as special issues of Nuclear Instruments and Methods in Physics Research, Section A [9].

Since 1977, these conferences were attended by foreign scientists, the representatives, practically, of all the leading SR centres the world around. Starting with 1982, there were 20 through 50 foreign participants in each.

On the basis of SSRC, three All-Union schools took place:

— EXAFS spectroscopy (1984);
— X-ray fluorescence elemental analysis using synchrotron radiation (1985), Synchrotron radiation — new possibilities of X-ray diffraction (1987).

The results of these schools were published in 3 monographs [10].

The last several years (especially 1991 and 1992) have been especially difficult for the arrangement of conferences and schools. Nevertheless, in 1993 two small international conferences were conducted on the basis of SSRC: "US-Russian Workshop on FEL Power Beaming", where the power beaming problems at satellites using FEL, and "Siberian HAZE-2" concerning the issues of applying synchrotron radiation and FEL to elemental analysis of aerosols. In July of 1994, the Russian conference SR'94 is scheduled to be held.

7. The analysis of the SSRC activities shows that this kind of management allows one

— to exploit effectively unique expensive facilities — SR sources;
— to unite the efforts of many institutes and institutions in the creation of the needed experimental equipment;
— to effectively exchange by experimental teams from various institutes and institutions;
— to arrange the training not only for some students and post-graduate students, but for teams from the other SR centres;
— to arrange the fruitful cooperation between science and industry;
— to effectively exploit international cooperation.

References

[1] Synchrotron radiation in the INP of the Siberian Division of the USSR Academy of Sciences (bibliography): Part 1 (1986). Part 2 (1988). Part 3 (1990).

[2] Kulipanov G.N. (1981) (Project of SR sources including research and development for insertion devices in the USSR. NIM, Sec. A, v. 303, №3, pp. 435—448.

[3] Kulipanov G.N., Mezentsev N.A. *et al.* (1992) Development of superconducting compact storage rings for technical purposes in the USSR (invited). Rev. Sci. Instrum. 63(1), pp. 731—736.

[4] Gavrilov N.G., Gorniker E.I. *et al.* (1991) Project of a race-track microtron-recuperator for free electron laser. IEEE J. Quant. Electron., v. 27, №12, pp. 2626—2628.

[5] Rather J., "NSA's perspectrive on laser power beaming", Proceedings of OE LASE'92, SPE Conference, 1628.

[6] Report on SR works in the INP of the Siberian Division of the USSR Academy of Sciences. (1981). p. 198. Novosibirsk

[7] Proceedings of the 5th National Conf. on SR Utilization: SR-82. (1982) p. 472. Novosibirsk

[8] Proceedings of the 6th National Conf. on SR Utilization: SR-84. (1984). p. 350. Novosibirsk

[9] Proceedings of the National Conf. on SR Utilization (Publ. at NIM): 7th, SR-86 — Nucl. Instrum. and Meth. (1987). v. A261, №1/2; 8th, SR-88 — Nucl. Instrum. and Meth. (1989). v. A282, №2/3; 9th, SR-90 — Nucl. Instrum. and Meth. (1991). v. A308, №1/2.

[10] Collected books:

X-ray method for study of amorphous body structure: EXAFS Spectroscopy / Inst. of Catalysis. (1988). Novosibirsk: Nauka, p. 306.

Synchrotron radiation in geochemistry: Sci. work collection / Inst. of Geology & Geophysics. (1989). Novosibirsk: Nauka, p. 151.

Synchrotron radiation diffractometry. (1989). Novosibirsk: Nauka, p. 144.

♦

INTERNATIONAL CENTRE FOR CATALYST CHARACTERIZATION AND TESTING: TWO YEARS OF ACTIVITY

K.I. ZAMARAEV
International Centre for Catalysts Characterization and Testing, Boreskov Institute of Catalysis of SB RAS, Novosibirsk, Russia

J.M. THOMAS
Royal Institution of Great Britain, London, UK

J.H. BLOCK
Fritz-Haber-Institut der Max-Planck-Gesellschaft, Berlin, Germany

ABSTRACT. This report presents the most important stages in the formation of the International Centre for Catalyst Characterization and Testing, spheres of its activity relating to catalysis fundamentals. Research work in the Centre is supported and provided by Boreskov Institute of Catalysis, which is well recognized by the world scientific community. The pioneer results are reported, which were obtained for particular catalytic systems studied within the first joint projects with the universities, research institutes and companies of the USA, UK, Germany, Spain, etc. The plans for the future activities are summarized as well.

Keywords: joint project / fundamental studies / catalyst characterization / reaction mechanism.

Introduction

Catalysis is of key importance for the design of new technologies for chemical and petroleum industries. About 90% of chemical products are produced nowadays with the use of catalytic technologies. Modern studies of catalysis are carried out with the use of new sophisticated and very powerful instrumentation. They can be only done by specially trained personnel and are rather expensive. For rapid and reliable characterization of new catalysts internationally accepted methods for the control and certification of their properties are required.

Since 1971 the Coordination Centre for Development of New Industrial Catalysts and Improvement of Catalysts Used in Industry of member countries of the Council for the Mutual Economic Aid had been functioning at the Institute of Catalysis (Novosibirsk). By now the Centre has been transformed into the International Centre for Catalyst Characterization and Testing. The Centre characterizes and tests catalysts on the basis of internationally accepted standards, as well as develops new standards for this purposes. The Centre organizes and

V A Koptyug and J Klerkx (eds), Science Policy New Mechanisms for Scientific Collaboration between East and West, 141–152

carries out projects upon request from various organizations of Russia and foreign countries, as well as joint bilateral and multilateral projects.

Objectives

The Centre provides facilities for comprehensive research in the area of basic and applied catalysis, organizes and holds international projects in the field of catalysis.

Main Directions of Activity

The main directions of the Centre's activity are:

- comprehensive characterization and testing of various catalysts according to internationally accepted standards, including characterization of their chemical and phase composition (in particular, the composition of surface and subsurface layers), point and extended defects, structure of surface species, electronic state of surface atoms, surface area and porous structure, interaction between the active component and support, testing of catalytic activity, etc.;
- comprehensive study of particular commercial catalysts and processes with the aim to provide recommendations concerning their improvement;
- carrying out joint studies with Russian and foreign partners for the development of new catalysts, processes and technologies, as well as for search of new catalytic systems;
- development of standardized equipment and methods for characterization and testing of catalysts, and certification of their physico-chemical and operational properties;
- development of quantum chemical methods for catalyst study together with a package of computational programmes, and the use of these methods and programmes for the interpretation of spectroscopic data;
- development of methods of mathematical modelling of catalytic reactors and processes together with a package of computational programmes;
- organization of training and studies for foreign specialists within the scope of the Centre's activity;
- advertising and representation of foreign companies at exhibitions held in Russia and other regions.
- arrangement of exhibitions, seminars and conferences.

Organization of activity

In its work, the Centre uses the powerful potential of the Institute of Catalysis of the Siberian Branch of the Russian Academy of Sciences. This specialized institute is the biggest in the world working in this field. The Institute carries out fundamental and applied studies practically in all fields of catalysis: structural and mechanistic characterization at the molecular level of various catalysts and catalytic reactions, chemistry of catalyst preparation, kinetic studies of catalysis, mathematical modelling and engineering of catalytic reactors and processes, development and commercialization of new industrial catalysts and processes. Sophisticated instrumentation for *in situ* analytical control of catalyst state and reaction mixture composition, including instrumentation for adsorption, kinetic, structural and spectral studies is available at the Institute. All this allows to obtain within short time intervals unique combination of data about the structure of active centers, detailed reaction kinetics and mechanism, and to use these data for purposeful design of new catalysts and catalytic technologies.

At present the Centre carries out joint projects with universities, research institutes and industrial companies of the USA, Germany, Great Britain, Spain, Italy and India.

International Advisory Board

The International Advisory Board of the Centre consists of the following specialists of the world renown.

1. Zamaraev K.I.	Boreskov Institute of Catalysis, Chairman, Russia
2. Bal'zhinimaev B.S.	Boreskov Institute of Catalysis, Russia
3. Bell A.T.	University of California, USA
4. Block J.H.	Fritz-Haber-Institut, Germany
5. Gryaznov V.M.	Peoples' Friendship University, Russia
6. Lovink H.J.	Akzo Chemical International, the Netherlands
7. Moiseev I.I.	Kurnakov Institute of General and Inorganic Chemistry, Russia
8. Parmon V.N.	Boreskov Institute of Catalysis, Russia
9. Rao C.N.R.	Indian Institute of Science, India
10. Somorjai G.A.	University of California, USA
11. Tamaru K.	University of Tokyo, Japan
12. Thomas J.M.	Royal Institution of Great Britain
13. Van Santen R.A.	Eindhoven University, the Netherlands
14. Vedrine J.C.	Institut des Recherches sur la Catalyse, France
15. Zhidomirov G.M.	Boreskov Institute of Catalysis, Russia

Management

Scientific supervisor of the Centre is Academician Kirill I. Zamaraev, Director of the Boreskov Institute of Catalysis.

Executive Director of the Centre is Professor Bair S. Bal'zhinimaev.

First Results of Activity

As has already been mentioned, the Centre carries out joint projects with the universities, research institutes and companies from the USA, Great Britain, Germany, Spain, Italy and India.

In this section some results are presented that were obtained within the framework of those projects.

(A)

Sponsors:

United Nations
Economic Commissions for Europe, Latin America, the Caribbean, and Africa.

United Nations
United Nations Environment Programme

B.I.R., Bureau International de la Récupération

The International Solid Waste and Cleansing Association

Commission of the European Communities
Directorate-General XII

Russian Academy of Sciences
Institute of Catalysis

International Union of Pure and Applied Chemistry

European Society of Membrane Science Technology

THE CLUB OF ROME

and 47 international trade associations and professional societies representing industry, trade, research and development, agencies related to recycling and the sections of the conference.

Printed on recyclable paper

Second Circular and
Final Call for Papers

ReC'93

International Recycling Congress
at the Re'93 Trade Fair
Chemicals, Oil and Solvents Section

19-22 January 1993
PALEXPO
Geneva, Switzerland

(B)

Dec. 31. 1990,
Oil & Gas Journal

Siberian catalyst research center works at world class level

Herman J. Lovink
Akzo Chemicals
International bv
Amersfoort, The Netherlands

The Soviet Union's Novosibirsk catalysis research and development center is one of the largest such facilities in the world. It is in the heart of Siberia on the Ob River some 2,000 miles east of Moscow.

The center's original charter was for fundamental research, but in the past 5 years the scientists there have been looking for practical uses for their findings.

The scientists, to name a few, are Profs. Zamaraev, Parmon, and Ione, and Dr. Startsev. They and their staffs are well known in their fields. They serve on the boards of several international journals.

The fact that Soviet institutes are state-owned means that their results go to difficult-to-trace patents and journals. Perestroika means, however, that these results should also be commercially applied in competition with everyone else.

A new group, called "Katalyzator," is trying to commercialize the findings from several institutes, plants, and other laboratories. A whole spectrum of new processes and catalyst improvements in specialty chemicals is now being offered to the Soviet industry as well as to the West.

Young managers, more fluent in English, are pushing new developments towards commercialization. With time, we will see more coming out of this working group. Fig. 1 indicates its objectives.

Modern ways of marketing and the know how to do it are in keen demand in the Soviet Union. Opening up to the world outside is of prime importance for Novosibirsk because Siberia is still far away and rather isolated.

Russia is big and so are the number of institutes and persons engaged in research and development work. Some 2,000 are working at Novosibirsk. Four hundred of them have academic degrees. At present the ratio between fundamental research and applied studies (pilot plant scale) there has been brought down to about 70 fundamental to 30 applied. This ratio in the West is

Fig. 1 A and B.
Illustrations world recognition of the work carried at BIC.

Project 1. Mechanistic Study of Butanols Dehydration on H-ZSM-5 Zeolites and Amorphous Aluminosilicate.

This is a joint project of the Boreskov Institute of Catalysis — BIC (Novosibirsk, Russia) and the Royal Institution of Great Britain — RI (London, United Kingdom). The project leaders are Prof. K.I. Zamaraev and Prof. J.M. Thomas. The project is supported by the RAS, the Siberian Branch of RAS and the Royal Society of London. It is carried out via exchange of scientists between BIC and RI. Experimental work is made mainly at BIC. The project was started before the creation of the Centre and now is a part of the Center's activity.

The results are published or prepared for publication in scientific journals (Refs. 1a—k in the list of publications).

Figures 2—6 present some important experimental data obtained within this project. Figure 7 presents the proposed mechanism of butanols dehydration.

The obtained results are of general importance for the acid-base analysis on zeolites.

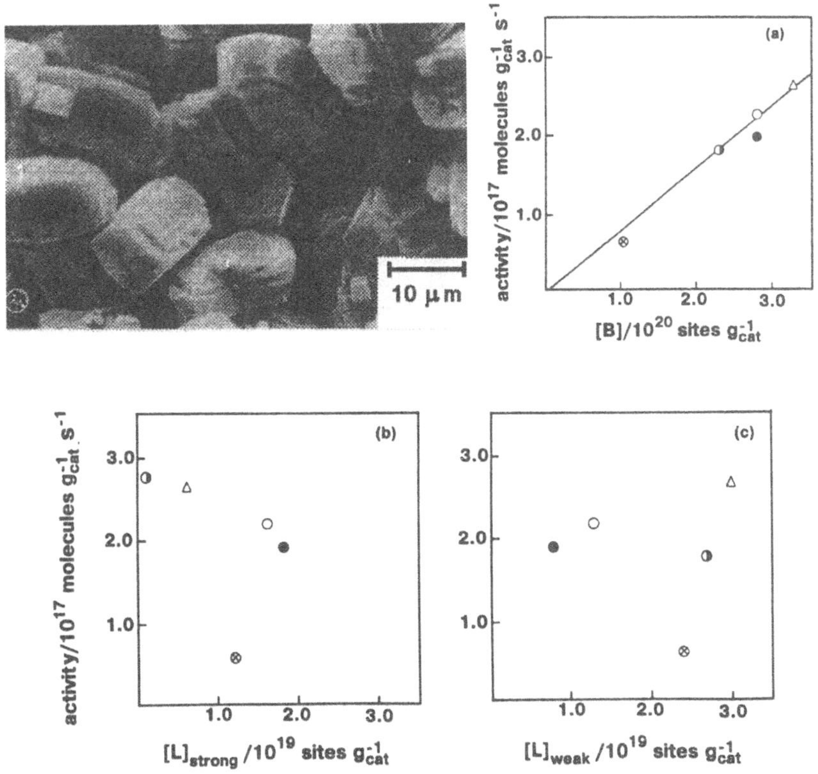

Fig. 2.

Scanning electron micrograph [1d] for H-ZSM-5 sample and relation [1b] between the activity of ZSM-5 and AAS in iso-butanol dehydration to butene at 397°K and the number of: (a) — Bronsted acid sites [B]; (b) — strong Lewis acid sites $[L]_{strong}$ and (c) — weak Lewis acid sites $[L]_{weak}$.

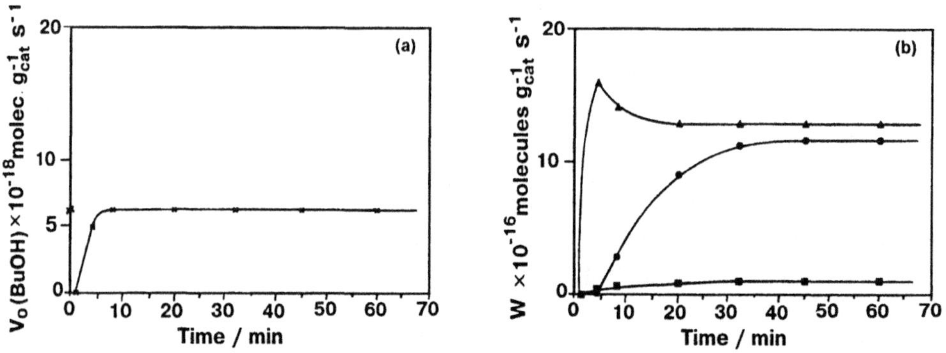

Fig. 3.
Kinetics of n-butanol consumption (a) and dehydration (b) in H-ZSM-5 (flow microreactor, 399°K); ▲ — water, • — di-n-butil ether, ■ — butene, x — unreacted n-butanol.

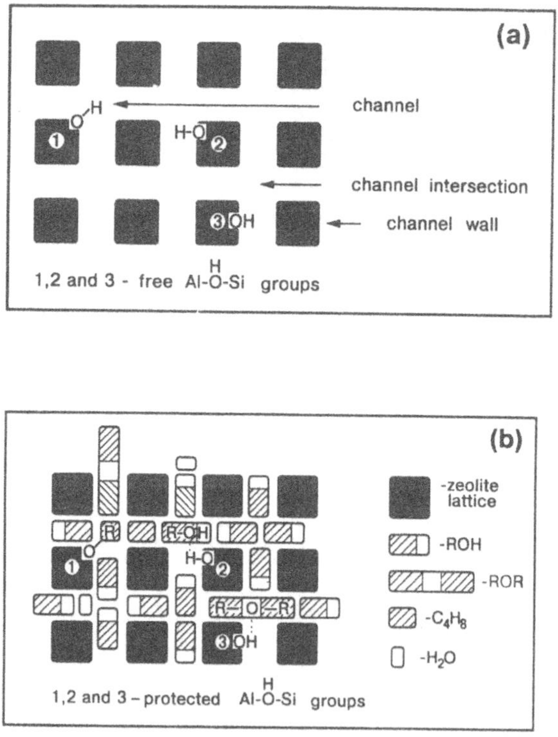

Fig. 4.
Schematic representation of H-ZSM-5 porous structure — (a) and of molecules of butanols of reaction intermediates and of products in H-ZSM-5 pores — (b). Polar and non-polar fragments of the molecules are shown as white and dashed areas, respectively.

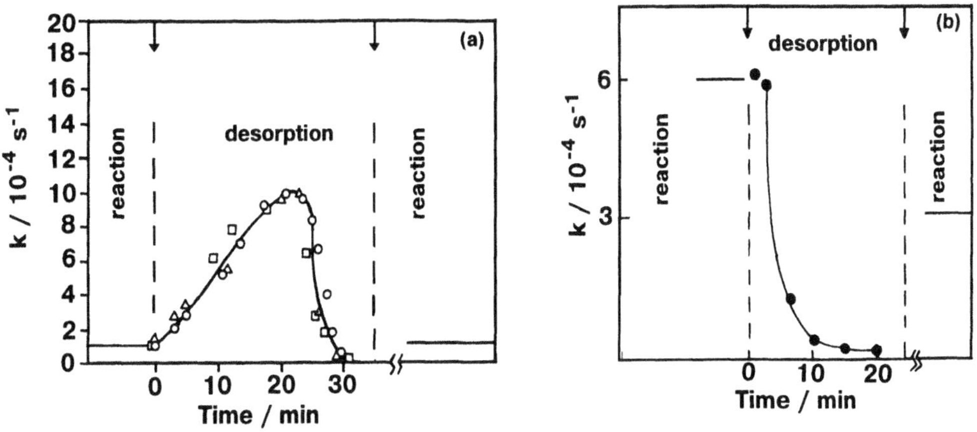

Fig. 5.
Rate constants of butanols dehydration into butene over H-ZSM-5 under steady-state conditions and subsequent desorption of butene when the helium + alcohol flow is switched to the pure helium flow:
(a) — sec-butanol at 369°K;
(b) — iso-butanol at 397°K.
The horizontal lines on the extreme right denote the reaction rate constants after returning to the previous helium + butanol flow.

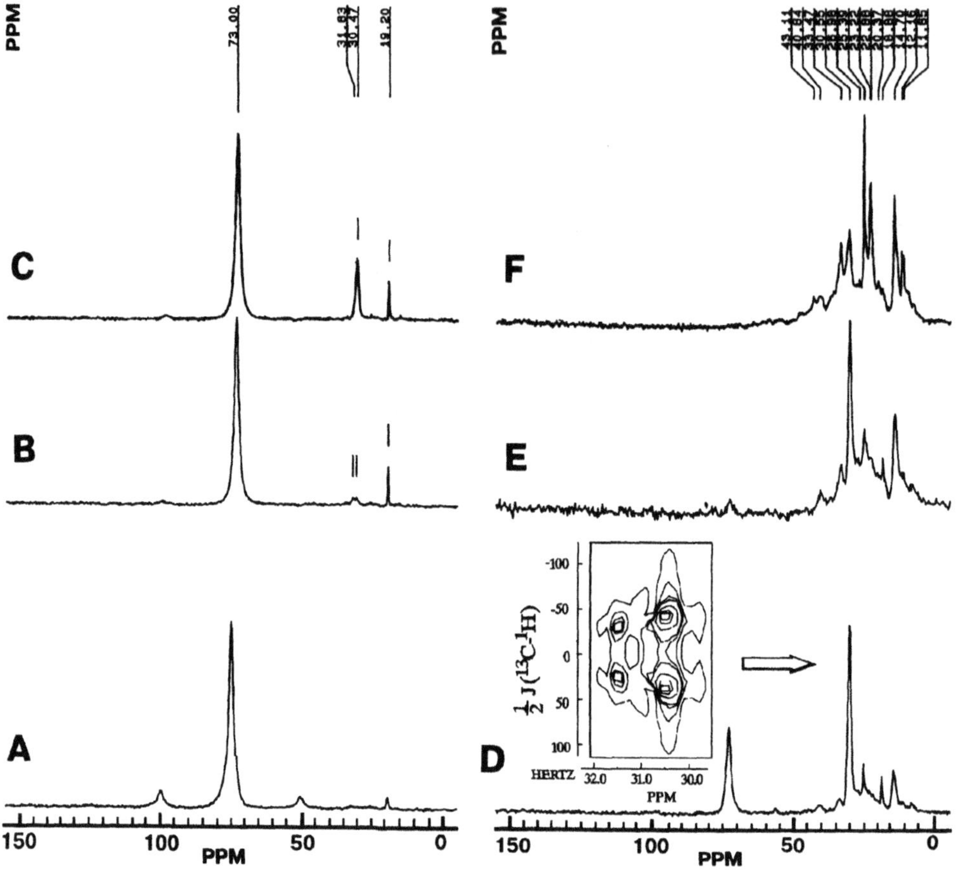

Fig. 6.

Variation of the ^{13}C CP/MAS NMR spectrum of $(CH_3)_2CH^*CH_2OH$
adsorbed in H-ZSM-5 upon successive heating of the sample for a certain time at various T:
(A) — after adsorption at 298 °K,
(B) — 70 min at 343 °K,
(C) — 40 min at 398 °K,
(D) — 60 min at 413 °K (inserted is the contour plot
of the 2D J-resolved ^{13}C MAS NMR spectrum in the vinicity of the signal at 30.5 ppm,
(E) — 60 min at 423 °K,
(F) — 60 min at 448 °K.

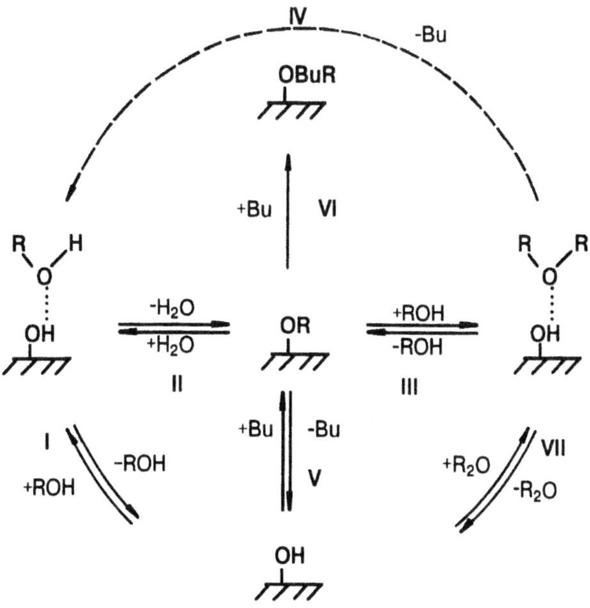

Fig. 7.
Mechanism of butanols dehydration.

Project 2. Oscillating Phenomena during CO and H_2 Oxidation on Platinum: Study at the Molecular Level.

This is a joint project of the Fritz Haber Institut — FHI (Berlin, Germany) and BIC (Novosibirsk, Russia). Project leaders are Prof. J.H. Block (FHI) and Prof. V.V. Gorodetskii (BIC). The project is carried out via the exchange of scientists between FHI and BIC. The experiments are carried out mainly at FHI.

The results are published or prepared for publications in scientific journals (Refs. 2a—d in the list of publications).

For example, Fig. 8 presents the oscillating phenomena that were observed at the atomic scale resolution with the help of the Field ion microscopy (FIM).

This is the first example of observing oscillation behaviour of reaction dynamics with the atomic scale resolution.

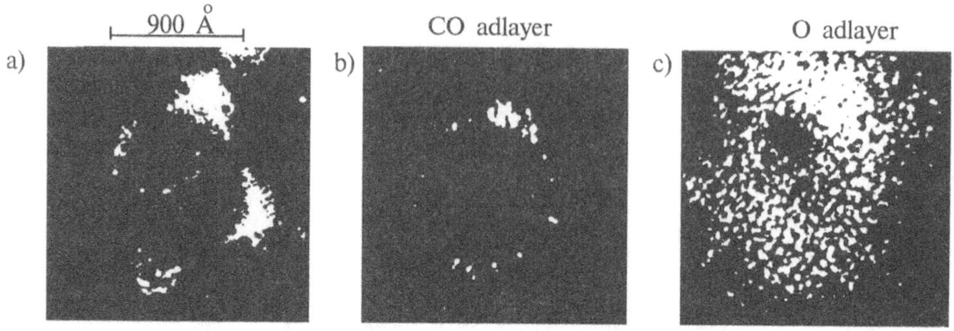

Fig. 8.
Field ion microscopic (FIM) images of the CO + CO_2 reaction oscillations on a (001) oriented Pt field emitter.
(a) Clean surface, FIM He/Ne at 16 kV, $p_{He} = 3 \times 10^{-4}$ mbar, $p_{Ne} = 8 \times 10^{-6}$ mbar, T = 78 °K, tip radius 850 Å.
(b) O_2^+-image, CO_{ad}-surface at 7.8 kV, $p_{O_2} = 5 \times 10^{-4}$ mbar, $p_{CO} = 5 \times 10^{-5}$ mbar, T = 300 °K.
(c) O_2^+-image, CO_{ad}-surface after "titration reaction".
Time of reaction 700 s at 300 °K at 7.8 kV, $p_{O_2} = 5 \times 10^{-4}$ mbar, $p_{CO} = 10^{-7}$ mbar.

Project 3. Development of the Open-Shell Self-Consistent Field Theory for Highly Symmetrical Systems.

This is a joint project of the BIC (Novosibirsk, Russia) and the Institute of the Computational Chemistry, University of Girona — ICCUG (Girona, Spain). Project leaders are Prof. B.N. Plakhutin (BIC) and Prof. R. Carbo (ICCUG).

A post-graduate student from the University of Girona J.M. Oliva is currently working at BIC under this project.

The paper "Open-Shell Hartree-Fock Energy for Molecular Structures of Icosahedral Symmetry" is prepared for publication [3].

The project will result in development of the theory that will allow to make much more reliable quantum-chemical calculations of the molecular structures with the open electronic shell which are important for catalysis.

Project 4. Basic research on the understanding of catalytic action of silver catalysts in ethylene epoxidation.

This is a joint project of the Boreskov Institute of Catalysis (Novosibirsk, Russia) and the Scientific Design Company (Little Ferry, USA).

Project leaders are Prof. B.S. Bal'zhinimaev (BIC) and Dr. V.S. Bhise (SDC).

Fig. 9 illustrates some features of the elucidated reaction mechanism at the molecular scale.

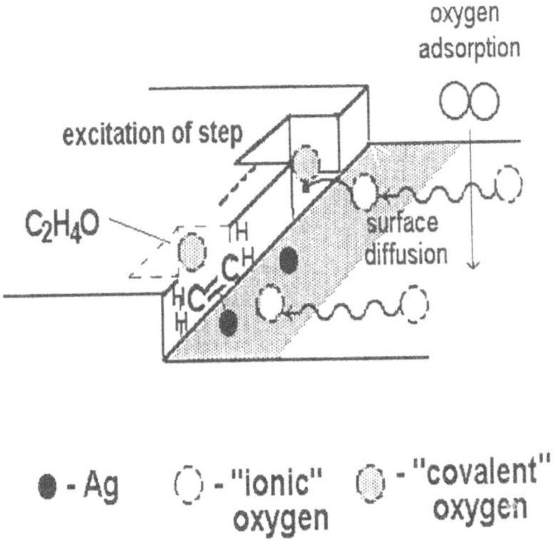

Fig. 9.
Mechanism of ethylene epoxidation on silver.

Plans for Future

In the future we plan:

1. To continue the ongoing projects with the universities, research institutes and companies from various countries.
2. To increase the number of joint projects with other institutions from Russia.
3. To start in 1994 six new projects with l'Institut des Recherches sur la Catalyse (Lyon, France).
4. To organize in 1995 an International Workshop on Standard Procedures and Reference Materials for Catalysts Characterization and Testing.
5. To organize regular annual meetings of the International Advisory Board of our Centre at Novosibirsk in 1994 and 1995.
6. To start selecting projects within BIC through the Centre in a competition with the proposals from others institutions from Russia and abroad.

Requests for support from NATO

We request for the support from NATO of the following activities:

1. Organization of two annual meetings of the International Advisory Board of our Centre in 1994 and 1995.
2. Organization of the International Workshop on Standard Procedures and Reference Materials for Catalysts Characterization and Testing in Novosibirsk in 1995.

References

[1a] Makarova M.A., Williams C., Romannikov V.N., Zamaraev K.I. and Thomas J.M. (1990). J. Chem. Soc. Farad. Trans., 86, 581.

[1b] Williams C., Makarova M.A., Malysheva L.V., Paukshtis E.A., Zamaraev K.I. and Thomas J.M. (1990). J. Chem. Soc. Farad. Trans. 86, 3473.

[1c] Makarova M.A., Williams C., Thomas J. and Zamaraev K.I. (1990). Catal. Lett., 4, 261.

[1d] Williams C., Makarova M.A., Malysheva L.V., Paukshtis E.A., Talsi E.P., Thomas J. and Zamaraev K.I. (1991). Catal. Lett, 127, 377.

[1e] Stepanov A.G., Romannikov V.N. and Zamaraev K.I. (1992). Catal. Lett., 13, 395.

[1f] Stepanov A.G., Zamaraev K.I. and Thomas J.M. (1992). Catal. Lett, 13, 407.

[1g] Stepanov A.G., Maryasov A.G., Romannikov V.N. and Zamaraev K.I. (1993). Proc. 10th Intern. Congress. Catal., Budapest, 1992, Eds. Guszi L., Solyosi F. and Tetenyi P. Part A, p. 621, Akademiai Kiado/Elsevir, Budapest.

[1h] Stepanov A.G. and Zamaraev K.I. (1993). Catal. Lett, 19, 153.

[1i] Makarova M.A., Paukshtis E.A., Thomas J.M., Williams C. and Zamaraev K.I. to be submitted for publication.

[1j] Makarova M.A., Thomas J.M., Williams C. and Zamaraev K.I. to be submitted for publication.

[1k] Zamaraev K.I. and Thomas J.M. Reaction pathways, key intermediates and pore effects in dehydration of butanols over H-ZSM-5 zeolites and amorphous aluminosilicate (prepared for publication).

[2a] Gorodetskii V.V., Drachsel W. and Block J.H. (1993). Catalysis Lett., 19, 223.

[2b] Gorodetskii V.V., Block J.H., Drachsel W. and Ehsasi M. (1993). Appl. Surface Sci, 67, 198.

[2c] Gorodetskii V.V., Drachsel W., Ehsasi M. and Block J.H. (prepared for publication).

[2d] Drachsel W., Gorodetskii V.V. and Block J.H. (prepared for publication).

[3] Plakhutin B.N., Oliva J.M. and Carbo R. (prepared for publication).

♦

TOWARDS SUSTAINABLE DEVELOPMENT: CONTRIBUTIONS OF THE BORESKOV INSTITUTE OF CATALYSIS

H.J. LOVINK
Akzo Catalysts, Senior Technical Associate, Amersfoort, the Netherlands

ABSTRACT. The forecasts and ways of rational energy use in future is considered. Catalysts and processes on its basis exemplified by oil refining processes are one of the perspective tools in solving energy problem. As an example of efficient combination of the fundamental and applied studies in this field the Boreskov Institute of Catalysis is illustrated.

Keywords: forecast / energy / development / fundamental research / science policy.

A. Introduction

The 1992 UN Conference on Environment and Development in Rio de Janeiro made it clear that mankind cannot continue to use fossil fuels and develop as in the recent past. The increasing quantities of CO2, the decreasing natural resources and the extinction of many larger living species are warning signals that should not be ignored.

The growing awareness of these dangers already had alarmed the international scientific community, leading 15 years ago to a series of IUPAC conferences: "Chemicals Research Applied to World Needs", abbreviated CHEMRAWN. The 8th conference in this series was held in Moscow in September 1992.

B. Forecasts of energy use

Annually a few noteworthy institutions are preparing 2-year forecasts of energy use worldwide:

- The International Energy Agency - Paris,
- Department of Energy in Washington D.C.

Such forecasts based on extensive studies comprising the growths in Gross Domestic Product, population, industry, scientific and technological changes are useful for planners for detailed related areas.

Catalysts is one such section and Table 1 has been prepared by author H.J.Lovink on the basis of the average of a number of forecasts, plus a number of other correlations and relationships. The conclusion of these and other forecasts is that, for the time being, we will continue growing in numbers particularly in the transportation sector, although the "energy intensity" tons of oil equivalent per $ 1000 gross domestic product are going down already because of higher efficiencies and further automation. Petrochemicals and pollution equipment are obviously more rapid growers than oil and primary energy.

V. A. Koptyug and J. Klerkx (eds.) Science Policy: New Mechanisms for Scientific Collaboration between East and West, 153–157

	1990	2000	2010	Sources **
World Population, millions	5270	6233	7223	"Energy in Europe"
Primary Energy m.t.o.e. *	7768	9129	11476	IEA-Paris 1993
Energy Intensity t.o.e./$ 1000 GDP	0.41	0.38	0.36	OECD-Paris
Motorvehicles on-the-roads x10^6	583	755	975	"Degussa", ACEA; AKZO
Airtraffic passenger Kms x 10^9	1600		~ 3500	OECD-Paris
Transportation fuels				
Motorgasoline x 10^6 ton/y	736	-	1100	IEA Paris 1993
Diesel x10^6ton/y	405	-	683	IEA Paris 1993
Jet fuel x 10^6ton/y	176	-	310	IEA Paris 1993
Marine bunker plus-diesel x 10^6 ton/y	~100	-	~ 140	OECD Paris 1993
Crude oil consumption m.t.o. *	3061	3500***	3950	Authors HJL
Conversion needed, % of crude oil	33	39	44	Authors HJL
Catalysts for petroleum refining billion $/year (all)	1.6	2.5	~ 3.5	Authors HJL

Other "growers"				
Petrochemical feedstock m.t.o.e. *	175	260	350	IFP
Automotive catalysts $ 10^9	1.4	3.5	~ 5.0	miscellaneous
Automation equipment (all sorts) $ 10^9	28	55	-	"Instruments & Automation" 12/'93

* = millions tons oil equivalent.
** = only the major ones; many more have been used for details.
*** = ample capacity for refining is currently under construction.
$ = 1990 - US dollars

Table 1.
Total world growth of energy consumption, transport and "Catalysts" 1990—2010.

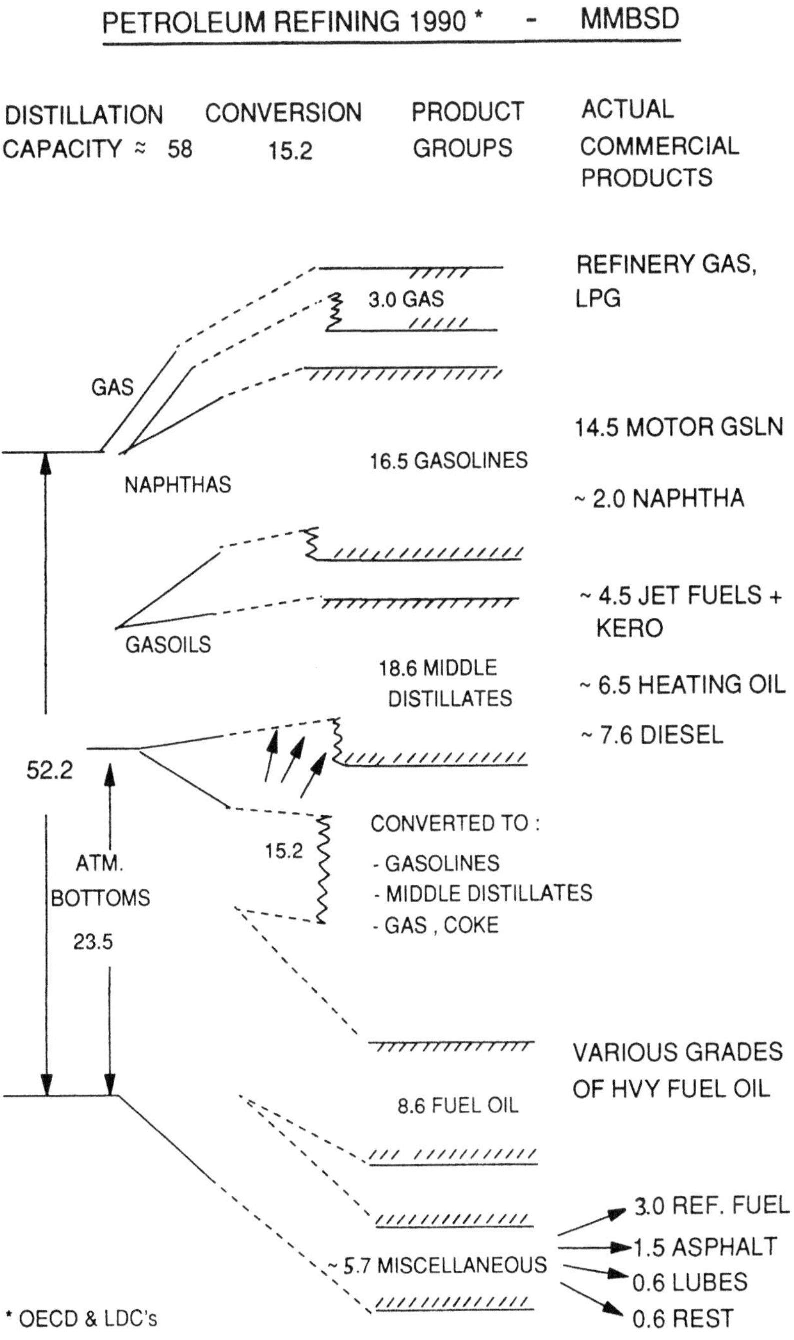
PETROLEUM REFINING 1990 * - MMBSD
DISTILLATION CAPACITY ≈ 58
CONVERSION 15.2
PRODUCT GROUPS
ACTUAL COMMERCIAL PRODUCTS
REFINERY GAS, LPG
3.0 GAS
GAS
14.5 MOTOR GSLN
16.5 GASOLINES
NAPHTHAS
~ 2.0 NAPHTHA
~ 4.5 JET FUELS + KERO
GASOILS
18.6 MIDDLE DISTILLATES
~ 6.5 HEATING OIL
~ 7.6 DIESEL
52.2
CONVERTED TO :
15.2
- GASOLINES
ATM. BOTTOMS
- MIDDLE DISTILLATES
23.5
- GAS , COKE
VARIOUS GRADES OF HVY FUEL OIL
8.6 FUEL OIL
3.0 REF. FUEL
1.5 ASPHALT
~ 5.7 MISCELLANEOUS
0.6 LUBES
0.6 REST
* OECD & LDC's

Fig. 1.

C. Conversion needs of the oil refining industry

As Table 1 demonstrates, the growth of the various oil products is not equal; residual oils for fueling industry, power, ships, etc. are staying behind relative to the light transportation fuels.

Figure 1 gives a quantitative picture of the year 1990; by 2000 and 2010 the conversion of heavy gas oils and residues will have to proceed further, meaning more use of catalytic processes and catalysts.

Longer term than 2010, the view on developments becomes less and less clear. By the year 2010—2050 new energy sources will be more urgently needed and no one knows what actually comes next.

Figure 2 has been taken from a booklet of the EC-Brussels "Energy in Europe 1992". It indeed suggests that now is the time to look ahead, because many years will be needed for the transformations that our society will need.

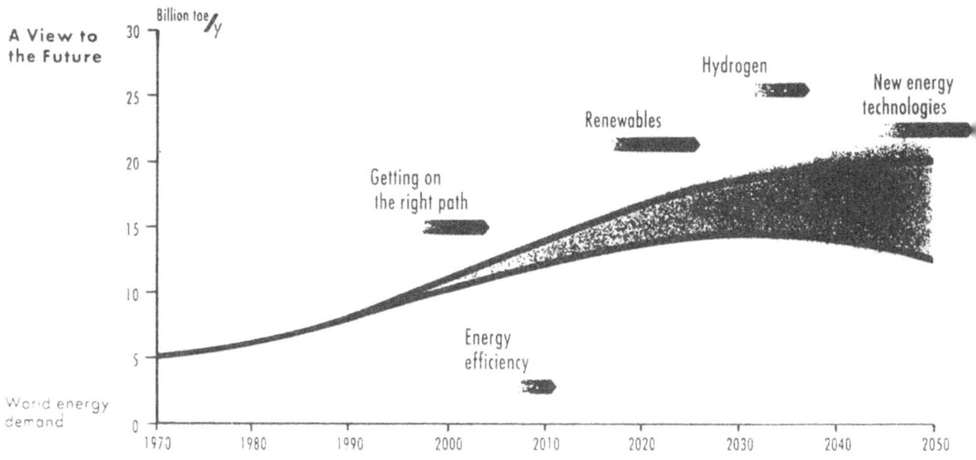

Fig. 2.

D. Science policy and fundamental research

Fundamental research is an inseparable forerunner and companion of new technological achievements; this holds particularly for catalysis and the added sketch is just a summary of the skills needed. (Fig. 3).

All this is available at the Boreskov Institute of Catalysis of SB RAS, Novosibirsk, the dozens of catalysts, carriers, zeolites, etc. The numerous new processes and hundreds of publications are witnessing this. Furthermore the increasing number of cooperation with Russian and foreign companies are keeping the work close to the mainstream of world attention.

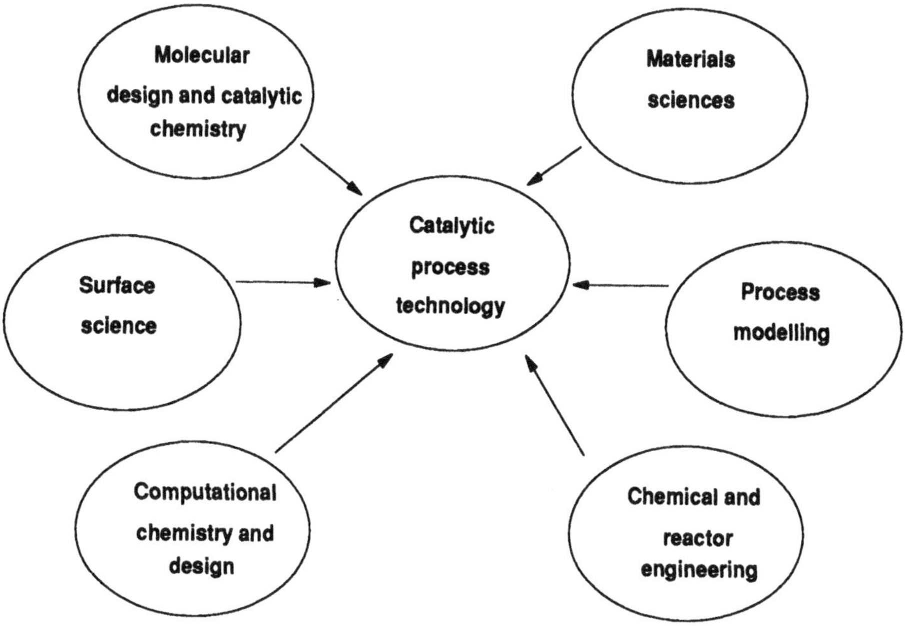

Fig. 3.

What the right science policy is, is a continuous effort of evaluation, careful consideration and gradual adjustments, as "Policy" in this area is a set of attitudes and actions aimed at achieving attainable goals together with all concerned.

Furthermore "Science Policy" looks always ahead:

— Far enough to see distant goals.

— Nearby in order to get things started in tandem with what our (industrial) society and its people can absorb and apply.

Alternative, sustainable energy sources and great savings in currently used fossil fuels and chemicals are on the horizon. However, much change and great efforts are still needed.

♦

MATERIAL SCIENCE PROGRAM OF THE SIBERIAN BRANCH OF THE RUSSIAN ACADEMY OF SCIENCES

F.A. KUZNETSOV
Institute of Inorganic Chemistry of SB RAS, Novosibirsk, Russia

ABSTRACT. Material science has a long tradition in Russia. From the very beginning of the Siberian Branch of RAS material oriented programs play a significant role in the activity of many Institutes. In recent years a number of integrated inter institute programs on materials were formed in different directions of material science: functional materials, construction materials, catalysts, synthetic minerals and gem stones, and others. The programs include all major steps of material investigation: material design, preparation of new precursors, new methods of materials synthesis, complex characterization of materials, construction and testing of prototype devices. The Siberian Branch has significant experience in technology transfer in the field of new materials and technologies. There are different ways of international cooperation in this area.

Keywords: new materials / functional / construction / materials / methods of material preparation / material characterization / material data banks / technology transfer.

Materials design, research and development are an old tradition of Russian science. To support this statement I may mention of Mikhail Lomonosov's study of glasses, D.I. Mendeleev's periodic table, N.S. Kurnakov's investigation of phase diagrams, A.F. Ioffe's contribution to semiconductor science,the Russian school of superconductivity (V.L. Ginzburg, A.A. Andreev and others).

From the very beginning of the Siberian Branch of RAS materials oriented projects constitute a significant part of the science program of many institutes of the Branch.

In the early 80s SB RAS considered and accepted an official program of research and development of materials as one of the priority programs of basic investigations of SB RAS.The official title of the program is "New materials and substances as a basis for creation of machines and technologies of new generation." The main objective of the program is to coordinate efforts of material scientists working in different institutes. The experience of cooperation proved to be useful in contacts with large industrial enterprises, in preparation of proposals for various grants.

The program includes all major topics of modern material science:

- Molecular design, search for new chemical compounds and development of new methods of their synthesis;
- Design and search of new materials;
- Methods and techniques of materials preparation;

V A Koptyug and J Klerkx (eds), Science Policy New Mechanisms for Scientific Collaboration between East and West, 159–162

— Development of new methods and instruments of material characterization;
— Development and permanent support of materials data banks, development of computerized data information systems, methods of modelling in material science and technology;
— Investigation of different possibilities of material application. Creation of prototypes of new solid state devices.

The major participants in the program are the following Institutes:
— Inorganic Chemistry (Novosibirsk)
— Solid State Chemistry (Novosibirsk)
— Physics of Semiconductors (Novosibirsk)
— Organic Chemistry (Novosibirsk)
— Organic Chemistry (Irkutsk)
— Physics (Krasnoyarsk)
— Thermophysics (Novosibirsk)
— Natural Sciences (Ulan-Ude)
— Carbon Materials (Kemerovo)

Many projects are being performed in close, every day cooperation with many other institutes of SB RAS:

Institutes of Catalysis, Nuclear Physics, Kinetics and Combustion, Geology and Geophysics and others.

The activity of SB RAS in materials is not restricted to the mentioned program only, which has to do mainly with so called functional materials. There are other material based programs on catalysts, new composite materials, synthetic minerals and gems and so on.

In this short presentation it is impossible to describe the content and results of the program, which involves a few hundred researchers. So, I will give only some examples of the investigation illustrating different parts of the program.

New Chemical Compounds

Participants of the program are involved in design, synthesis and research of a large number of compounds of different types:
— Silicon organic compounds;
— Coordination compounds of transition metals (Pt group elements, Au, Ni, Co, W, Mo...);
— Metal organic compounds;
— Boron hydrides, metal carboranes;
— Clathrate and cluster compounds;
— Rare earth sulfides, oxisulfides;
— Multi-component mixed oxide compounds;
— Fluorinated metal organic compounds and carboxiates.

Material Preparation Technique

All major known techniques for material preparation are presented in member institutes. Following are examples of the most original ones.
— Multiple versions of the chemical vapor deposition technique with use of inorganic and metal organic compounds, thermal and non thermal excitation (plasma, laser, UV and IR light enhanced processes);
— Mechanochemical synthesis of volatile compounds — precursors for CVD and different kind of materials;

— Wide variety of crystal growth techniques, including original low temperature gradient method, growth of phases with very high melting temperature;
— Original variety of molecular beam epitaxy.

Material Characterization

Altogether, institutes involved in material program of SB RAS are able to make any kind of material characterization:
— chemical composition, including determination of content, spatial distribution, chemical state of main components, impurities and dopants;
— crystal structure, structure of surfaces and interfaces in multi phase materials;
— electronic structure of materials and phases;
— phase composition of bulk and thin film materials;
— thermodynamic properties;
— optical, electric, magnetic and mechanical properties.

In addition to conventional characterization techniques SB RAS employs some unique methods and instruments like analytical stations with synchrotron radiation sources, X-ray spectroscopy of "soft range": 1—1000 Å, electrochemical microphase analysis, new variations of mass spectrometry, chromatography, resonance spectroscopy and others...

Material Data Bases and Modelling

A number of material oriented factual data bases have been developed and are permanently supported by institutes, participants of the program:
— Data bank on property of electronic materials (SMET, Novosibirsk Institute Inorganic Chemistry);
— Data system on high temperature superconductors, (Novosibirsk Institute Inorganic Chemistry);
— Data bank on molecular spectroscopy (Novosibirsk Institute Inorganic Chemistry);
— Data bases on volatile compounds and high pure substances (Novosibirsk Institute Inorganic Chemistry);
— Library of quantum chemical programs (Institute of Chemical Kinetics and Combustion);
— Data bank on catalysts and catalytic processes, (Institute of Catalysis).

All above mentioned data systems are recognized at the international level.

Some other material oriented data bases are in the process of being established (on synthesis of substances and materials, on chemical aspects of ecology and others).

Attempts are being made to organize material data base network, which will provide direct access to existing data systems and participation in further development for material groups in Siberia, other groups in Russia and foreign partners.

At present materials oriented groups are among those most damaged by the economic conditions of the country. A significant part of funds for research in the framework of the program were previously provided by high tech industries which are now in difficulties themselves (electronics, communications, defense) and as a result demand for sophisticated materials research from domestic industry is significantly reduced. Participants of the program recently established new contacts with foreign partners, including commercial contacts. Earnings coming from these partners are not sufficient to compensate the decrease in domestic funding. As we understand they could never be sufficient. To provide the necessary funds we are doing two things.The first is to persuade authorities to invest more in material science which is so decisive for technical progres. Second, to find ways to earn more than we can now from domestic sources. The only sure way to increase earnings is to help large industry to find a new way of operation: to master new production, to reconstruct domestic market and to enter seriously the international market of high technology products.

Many of high tech enterprises are located in Siberian cities and Institutes of SB RAS have traditional ties with these institutions. Many of the enterprises are now involved in the process of "conversion". In many cases enterprises are switching to the production of advanced materials and related items (high pure substances, new technology equipment, control and analytical instruments). A good example is the network of plants and industrial complexes belonging to the Ministry of Atomic Energy of Russian Federation (MAERF). Institutes of SB RAS cooperate with the enterprises in many areas of materials production:

— high pure substances (metals, non metallic elements, chemicals, metal organic volatile compounds);
— new purification processes and equipment;
— semiconductor materials;
— magnetic materials;
— materials for energy saving;
— ecomaterials;
— control and technological equipment and instruments.

Academic institutes are using their traditional ties with foreign partners to help national industry to enter into international cooperation. In doing this we understand that assistance to our industry is necessary not only in the interests of industry itself but to the interests of research institutes which have contacts with the industry. It is also of interest to foreign parties, for if Russian industry will come to the international market poorly prepared it will damage itself, the market and all participants in the market due to its large power and capacity.

To help with the "internationalization" of domestic materials industry we use different channels of international cooperation. One of effective ways is cooperation in the framework of the newly organized (from April 1992) Asia Pacific Society for Advanced Materials (APSAM). Thus, in August 1993 APSAM held an annual meeting in Russia near Novosibirsk. The program of the meeting included a special workshop on the problem of "conversion" of Russian large industry and perspectives of wide scale production of high pure substances, semiconductors, materials for energy and environmental protection. The workshop was well attended by representatives of large industrial enterprises in Russia. As a follow up of the meeting, a seminar was held in Tokyo with participation of representatives of the Russian Academy of Sciences and Russian Ministry of Atomic Energy. The next meeting of this nature is planning to be held in India in the first part of 1994. The problem for discussion at this seminar is science-industry cooperation in field of advanced materials.

The described activities were organized by the International Centre for Materials Research — Siberia (ICMRS), which is an association of major Institutes of SB RAS, participants of the mentioned program of SB RAS. This association undoubtedly can be very useful in establishing cooperation of ES material science groups with Siberian colleagues. It can also be used to reach many material oriented institutions and specialists in countries of the Asia Pacific Region.

♦

THE CONVERSION OF BASIC CHEMICAL TECHNOLOGIES INTO THE SOLID STATE PROCESSES

N.Z. LYAKHOV and V.V. BOLDYREV
Institute of Solid State Chemistry of SB RAS, Novosibirsk, Russia

ABSTRACT. This is a brief discussion of new possibilities provided by solid state chemistry in chemical technologies. It is proposed to substitute the conventional "wet" processes by solid state ones in order to avoid the formation of liquid wastes. This may be a key to the environmentally safe technologies in various industrial applications.

Keywords: solid state chemistry / dry technologies / wastes.

In the present-day chemical industry all of the currently available processes can be reduced to a relatively small number of widely used operations, despite the great diversity of final products. With the exception of gas and oil refining, in the great majority of cases, chemical production uses solids as a raw material and yields solid products. However, chemical transformations are carried out in solutions, the operations of dissolution, mixing, extraction being the most wide spread in chemical technologies. It is these operations that make a major contribution to the pollution of the environment by chemical industry like in hydrometallurgy, pharmaceutics, production of cellulose, electrochemical deposition, etc.

The increasing requirements for the ecological safety of chemical industries and a strong demand to decrease the negative effect on the environment compel scientists and engineers to expend big investments for designing technologies of second level, the only purpose of which is sometimes merely the detoxication of wastes.

In the actual practice of large plants, this classic way leads to the construction of new devices and even plants in order to realise the so-called "end-of-pipe" technologies.

This approach is schematically represented in Fig. 1, where traditional production yields gaseous, liquid and/or solid wastes. To solve the problem of safe landing of wastes, this requires, practically, three new plants.

It should also be noted that in chemistry the problem of solvents is very often directly linked with the problem of shortage of fresh water (if water is a solvent), or with the problem of fire hazard of chemical processes in the case of organic solvents.

Solid state chemistry has arisen and is progressing as a science to a great extent under the influence of the idea that some substances can be transformed into other without preliminary conversion of them into gases or solutions. The main objection against the wide use of solid state chemistry in practice has always been the comparatively low reactivity of solids.

V A Koptyug and J Klerkx (eds), Science Policy New Mechanisms for Scientific Collaboration between East and West, 163–166

However, it is just this direction where science has made considerable and successful advances for the last two decades. At present, in a number of publications it is shown that the rates of solid state reactions can be competitive with the rates of reactions in liquids and even exceed these. In many cases, these solid state reactions are highly adaptable to streamlined production such as, for example, technological combustion (self-propagating high temperature). Solid state chemistry has developed methods of controlling over a wide range the rates of thermal decomposition of solids, solid state synthesis, polymorphous transformations. It is not surprising, however, that the examples of application of these achievements in chemical engineering fall again in the field of intensification of the processes of dissolution, leaching, adsorption, i.e. the basic stages of traditional technologies. Such was the social order.

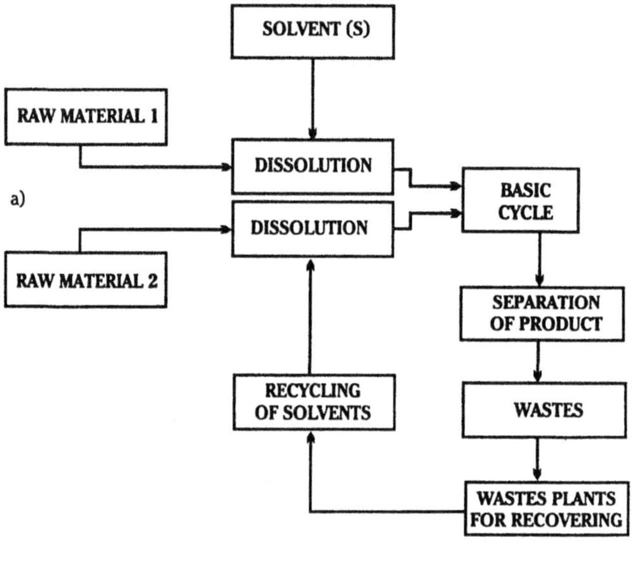

Fig. 1
Schematic representation of conventional ("wet") technologies (a) in comparison with solid state ("dry") ones (b).

At the same time, the progress in solid state chemistry enables the approach to the ecological problems of chemical industry to be radically changed. If one could manage to demonstrate the possibility of conversion operations in various branches of chemical industry to solid state processes, this would be an incitement to the application of the principle "not to produce wastes instead of controlling them". The so-called "dry" technologies have a lot of advantages over the current ones. Among these advantages, the main one is an intensive decrease or complete abandonment of the usage of solvents and hence the elimination of the associated operations of purification and inevitable hazardous pollutions.

The mechanochemical method of production of the soluble form of apatite can be cited as a classic example [1, 2]. The conventional method of production of phosphate fertilizers fully corresponds to the scheme of Fig. 1. In this case, sulphuric acid is used as a solvent and the preparation of granulated superphosphate (the solid product!) is accompanied by the production of a considerable amount of harmful solid and liquid wastes polluting the environment. The elegance of the mechanochemical method is in that the consumers' properties, i.e. the conversion of phosphates into the soluble state, are achieved practically in one stage. It should be noted that here again this stage, i.e. mechanochemical activation, does not essentially differ technologically from the operation of preliminary grinding of natural apatite used in the "wet" superphosphate technology. Mills should be replaced by special activators and all the rest including treatment by sulphuric acid becomes unneeded.

Unfortunately, the wide practical application of the methods of solid state chemistry is hindered by two circumstances.

First, the subject of investigations in solid state chemistry has not always been concerned with the processes directly meeting technological needs.

Second, the new methods of controlling the reactivity of solids, for example, the methods of mechanochemical activation are not sufficiently provided with technological equipment of sufficient productivity. This hinders the commercial application of techniques already devised and is a deterrent to the interest of potential consumers to the design of "dry" technologies.

A special international project proposed by the Institute of Solid State Chemistry in Novosibirsk aims to change the situation. The goal of the project is to estimate the prospects of conversion of basic chemical technologies to solid state processes. It is suggested that the estimation would be based on a careful comparison of the technical and economic indices of "dry" technologies with traditional ones, with regard to the expenditures needed to achieve comparable ecological effect. This should be preceded by special engineering studies and the design of pilot plants of demonstration type in order to draw the attention of engineers and designers to the potential advantages of the methods used in solid state chemistry.

This does not means that the conversion from the existing production processes to the new solid state ones should be carried out simultaneously and everywhere. In many cases, the introduction of even one solid state stage radically changes the process scheme, the proportion of energy expenditure and capital investment, sharply decreases the amount of harmful pollutants per unit of the finished products. The non-palladium technology of metallization of holes on printed circuits is a good example [3, 4]. The substitution of the "wet" chemical copper plating to the "dry" method of preparation of the primary layer of copper based on the thermal decomposition of copper salt allowed one not only to shorten twice the processing, but also to decrease by a factor of 10^2 the specific effluents of copper, the termination of the process being the same, i.e. electrochemical deposition. It is just this aspect of the project that appears to be the most important, i.e. experts shifts in chemical engineering due to wider application.

References

[1] Boldyrev V.V. (1993). Solid State Chemistry and cleaner chemical Technologies. Chemistry for sustainable development. v. 1, №1 pp. 125—129.

[2] Paudert R., Harenz H., Hernicke G., Dunkel L., Pothig R., Kolosov A.S., Tschaikina M.V., Taranzova M.I., Gordeeva G.I. (1979). Zur Erhohing der Reactivitat von apatitischen Rohphisphaten durch mechanische Aktivierung. Z. anorg. allg. Chem. 452 141—150.

[3] Lomovsky O.I., Revzin G.E., Boldyrev V.V. (1991). Environmental Aspects of Dielectric Technology, Using Termal Decomposition Reaction. Zhurnal Vsesoyuznogo Obschestva Mendeleeva, №3 pp. 337-342 (in Russian).

[4] Lomovsky O.I., Boldyrev V.V. (1989). Non-Palladium Metallization in Printed Circuit Technology. Russian Journal of Applied Chemistry v. 62 №11, pp. 2444-2455 (in Russian).

♦

SPACE AND ENVIRONMENTAL RESEARCH AT THE SB RAS FACILITIES

G.Ya. SMOLKOV and G.A. ZHEREBTSOV
Institute of Solar-Terrestrial Physics of SB RAS, Irkutsk, Russia

ABSTRACT. This report gives an outline of the main areas of research pursued at the ISTP, its experimental bases, facilities and kinds of observations, results of scientific cooperation with scientists from NATO countries, and proposals for a cooperative solution of current problems with the NATO Scientific Committee's assistance.

Keywords: solar-terrestrial physics / Siberia / unique instruments / international collaboration.

The Institute of Solar-Terrestrial Physics (ISTP, before 1992 SibIZMIR - Siberian Institute of Terrestrial Magnetism, Ionosphere and Radio Wave Propagation) at the Siberian Branch of the Russian Academy of Sciences (SB RAS) was established in 1960. Its major directions of research are:

- solar physics and solar-terrestrial relationships;
- the Earth's magnetosphere and electromagnetic field;
- the upper atmosphere, ionosphere and ionospheric propagation of radio waves.

These cover virtually all the diversity of solar-terrestrial relationships (Fig. 1 a, b). During the last several years the Institute has initiated and accomplished work on the upper atmospheric and near-terrestrial space ecology. Therefore, the program for observations of solar activity, cosmic rays and geophysical phenomena is rather extensive. The observations are being carried out at target-oriented observatories located in a vast region at middle and high latitudes. Of the greatest value here are longterm sequences of observations (for example, those of geomagnetic disturbances cover more than a hundred years, ionospheric data - more than 40 years, and solar activity - more than 30 years), the unique instruments (Siberian solar radio telescope (SSRT), Large solar vacuum telescope (LSVT), cosmic-ray spectrograph (CRS), infrasound installations, and others), and the original devices (birefringent filters, magnetographs, and others). (See Table 1).

As a consequence of the strongly expressed individual characteristics of the phenomena and processes under study, their nature has not yet been fully understood, and most of the kinds of observations are being carried out in the monitoring mode with original or standard programs. The scientific and practical significance of the STP monitoring was shown, for example, in [1, 2]. In parallel with the above studies, original investigations are being carried out in all research areas, including laboratory modellings of plasma processes. Scientific instrument

V A Koptyug and J Klerkx (eds), Science Policy New Mechanisms for Scientific Collaboration between East and West, 167–176

making is well developed at the Institute. All the unique instruments, installations and original devices have been developed and constructed largely at the Institute.

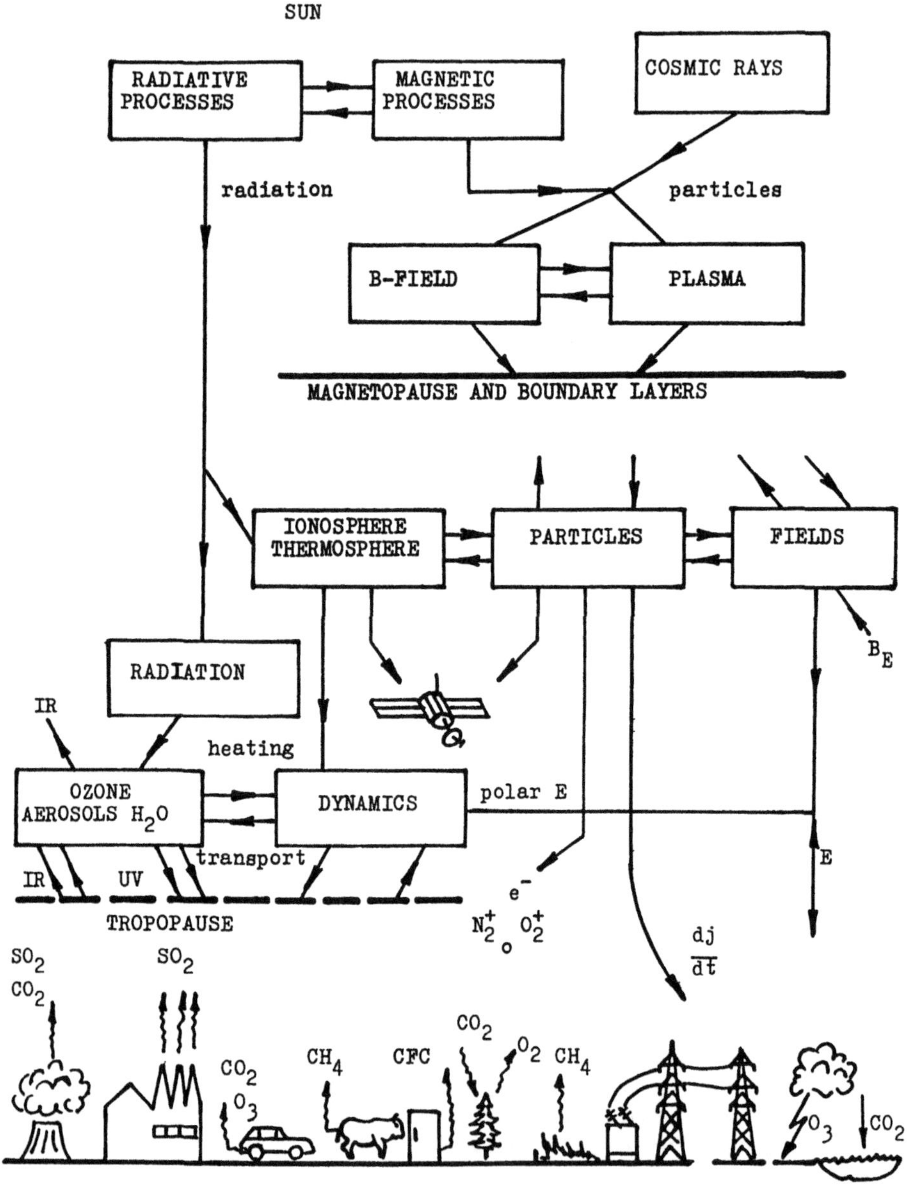

Fig. 1 (a)
Main channels of interaction between principal regions, composing the solar-terrestrial system.

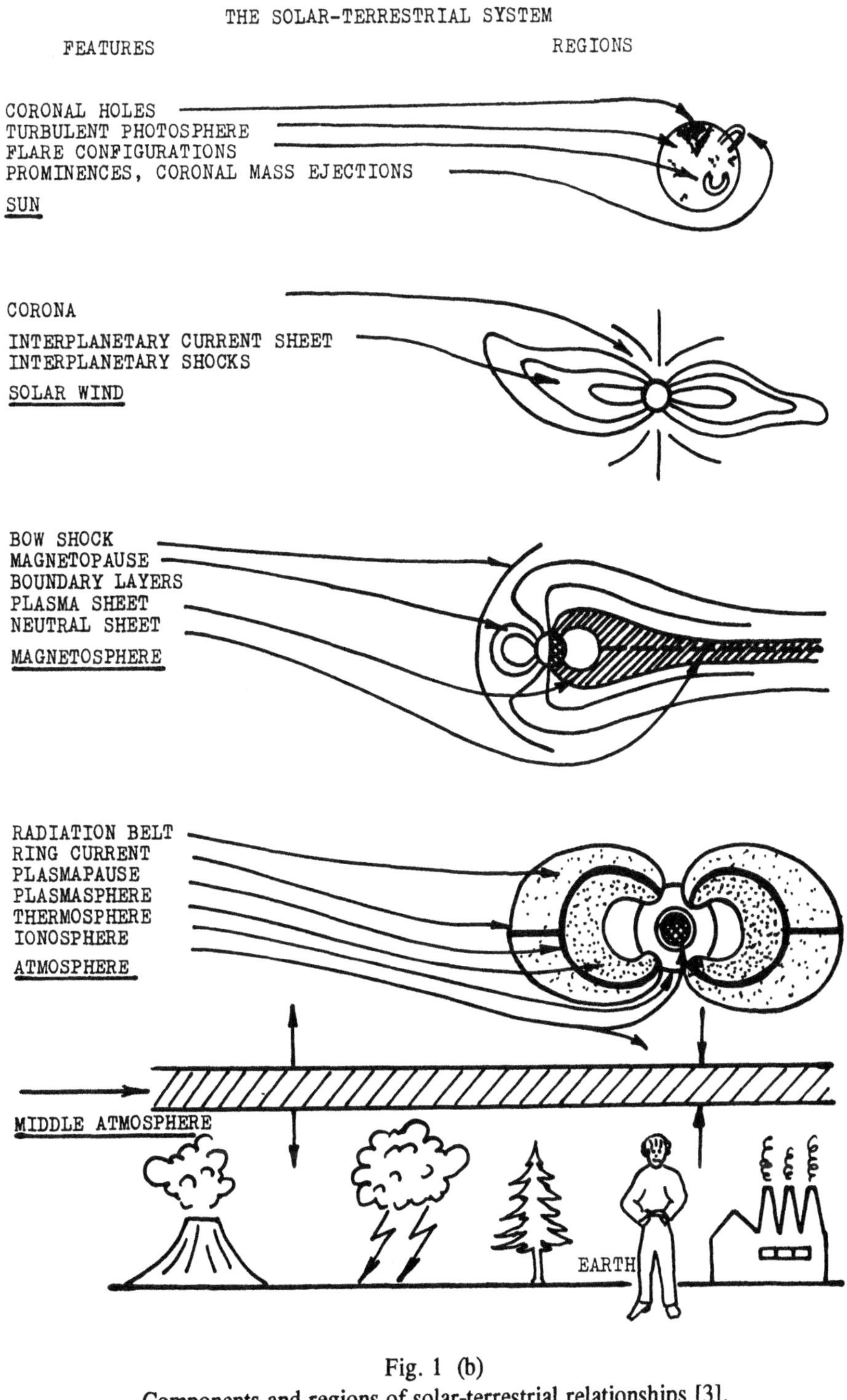

Fig. 1 (b)
Components and regions of solar-terrestrial relationships [3].

NN	Name, Instruments Facility	Coordinates Longitude E	Latitude N	Altitude above sea level, m	Kinds of observations, scientific objectives
1	Sayan Mountain Observatory: original automated solar telescope (AST), original solar telescope for operative predictions (STOP), two large unique coronographs (Fig. 2), chromospheric telescope, two stellar telescopes.	100°50' Cosmic-Ray Spectrograph (CRS)	51°40'	2012 400 2000 3000	a) Magnetic fields of active regions, prominences, of the Sun as a star, and background magnetic fields. Global oscillations and differential rotation. Spectroscopy of sunspots, flares and corona. Patrol of chromospheric flares and magnetic fields in sunspots. b) Monitoring of cosmic-ray intensity variations.
2	Baikal Astrophysical Obs.: original large solar vacuum telescope (LSVT), three original chromospheric telescopes (full disk in H-alpha and K Ca II, large-scale image)	104°50'	51°45'	500	Granulation, fine structure of the chromosphere, solar flares, spectroscopy of sunspots, large-scale solar activity.
3	Radioastrophysical Observatory: Unique Siberian Solar Radio Telescope (SSRT) (Fig. 3), two original small baseline radio interferometers (RIMB), radio polarimeters for cm and dm wavelengths, radio spectrograph 70—140 MHz, and photoheliograph.	102°10' Infrasound metering station: high-sensitivity wide-band infrasound installations.	51°50'	831	a) Solar activity monitoring since 1981; active regions, solar flares, oscillatory phenomena, energy release in the solar corona, filaments and coronal holes. b) Background infrasound monitoring to diagnose the influence of anthropogenic factors upon its generation regime.
4	Magnetic Observatory: Standard and unique magnetometers	104°24'	52°06'	500	Monitoring of variable geomagnetic field variations and disturbances in the range 0—5 Hz since 1968.
5	Baikal Magneto-Telluric Obs.: Thee-component induction nanoteslameter, telluric current installation, vertical-component magnetometer	107°18'	53°24'	460	Monitoring of short-period oscillations of the Earth's natural electromagnetic field and telluric currents since 1963.

Table 1.
The ISTP observatories.

6	High-Latitude Complex Geophysical Obs.: automated ionospheric station (AIS), installation for measuring radio wave absorption in the ionosphere, riometers, standard and unique magnetometers, automated cosmic-ray station, complex of auroral cameras and photometers	88°06'	69°24'		Monitoring of ionospheric and geomagnetic disturbances, radio wave absorption, cosmic-ray intensity variations, and auroras.
7	Complex Radiophysical Obs.: AIS for 1—18 MHz, AIS-017 (Vertical) for 1—19.9 MHz, vertical and oblique-incidence ionospheric sounding station "Bazis" for 0.3—19.9 MHz, radio transmitters AIS SP-3 for investigating the ionosphere and ionospheric propagation of radio waves, neutron supermonitor NM-64 as part of the cosmic-ray spectrograph (see № 1)	104°02'	52°27'	430	Monitoring of ionospheric disturbances since 1947, cosmic-ray intensity variations since 1957.
8	Ionospheric Obs.: installations for horizontal drifts in the lower ionosphere, phase variations of the ionosphere-reflected signal, ground atmospheric pressure	102°40'	51°50'	830	Experimental study of the dynamical regime of the lower ionosphere.
9	Observatory for radiophysical diagnostic of the atmosphere: a) radar unit "Dnepr" for 154—162 MHz	103°	53°	400	a) Ionospheric research by incoherent scatter; updating to IMST-radar is planned.
	b) chirp-sounder in the backscatter, oblique and vertical-incidence mode for 1—30 MHz, complex of azimuthal and scanning photometers, 8-channel frequency Doppler shift recorder	103°	51°42'	530	b) Study of mechanisms for ionospheric propagation of radio waves through backscatter, oblique and vertical-incidence sounding as well as density inhomogeneities of the neutral and ionized components of the upper atmosphere.

Table 1 (cont.)
The ISTP observatories.

Fig. 2.
The Large Coronograph of the Sayan Mountain Observatory
(view from the east, the objective lens 535 mm in diameter, focal length 8 m).

Fig. 3.
The Siberian Solar Radio Telescope of the Radioastrophysical Observatory
(view from the south, a 256-antenna crossed radio interferometer,
angular resolution 17", time resolution 7 msec).

In practice, the following activities are pursued:

— the provision of helio-geophysical data for predicting the geophysical situation as well as radiation conditions in near-terrestrial space;
— coordinate-trajectory monitoring of spacecraft, and spacecraft photometry and polarimetry;
— assessment of the near-terrestrial space environment pollution;
— studies of the environmental effects on spacecraft;
— diagnostics of the ecological situation in the upper atmosphere;
— determination of mechanisms, by which helio-geophysical and meteorological factors influence the organism of a healthy and a sick man (in collaboration with medical institutions since 1993).

The global character of the manifestation of solar-terrestrial relationships has motivated the involvement of the Institute as a whole or its observatories into the International cooperative arena. This is accomplished both under agreements with foreign observatories, universities and institutes and within International projects (such as STEP [3], Global Change, and others).

Our scientific relations with scientists from NATO countries and Western Europe were established and are maintained through personal contacts during International symposia and programs, mutual visits, the implementation of joint projects, bilateral agreements, participation in the implementation of intergovernmental agreements, participation in competitions of research projects of various scientific foundations (ISF, ESO, NATO, Association for Promotion of Science in the Independent States of the Former Soviet Union, and nominal foundations), and the creation of the International Centre for Solar-Terrestrial Physics on the basis of the ISTP.

They are based upon a mutually beneficial complement and pool of effort to solve common scientific problems, an exchange of observational data, publications, visits and experience, the joint use of unique instruments and facilities, reciprocal information, and assistance with the participation in international activities as well as visits.

Thus, cooperative investigations of active regions of the solar atmosphere with the University of Athens (Greece) using data from the SSRT with adaptation of previously accumulated experience and the software for solar radio image synthesis made it possible to detect microwave emission sources above the magnetic field polarity inversion line [4], indicative of energy storage and flare buildup. Progress in this direction can be secured by a grant from the Association for Promotion (see above), with the involvement of observatory Arcetri (Italy).

By modernizing the SSRT receiving system in cooperation with the Institute of Applied Physics of the University of Berne (Switzerland), it was possible — by improving the time resolution to 7 ms — to embark on an investigation of fast processes during flares and to come substantially closer to the study of events of primary energy release and charged particle acceleration (fulfillment of a long-standing dream of astrophysicists) [5].

In collaboration with the Observatory Weissenau of the University of Tuebingen (Germany), a study was made of characteristics of the solar flare radio emission [6].

A nonlinear one-dimensional dynamo model was developed by doing joint research with the Institute of Astrophysics (Potsdam, Germany) into nonlinear dissipation mechanisms for large-scale magnetic fields in turbulent media [7].

Through the implementation of a joint project with the Space Physics Institute (Sweden) a study is being made of the main large-scale (global) physical processes that are responsible for substorms and storms in the Earth's magnetosphere and the related electric currents on the basis of the ISTP magnetogram inversion technique and of Viking spacecraft data of measurements from the Space Physics Institute [8a, b].

In cooperation with the University of Illinois (USA) a study was made of small- and large-scale ionospheric irregularities using satellite navigation signals [9].

Intentions have been defined in regard to joint investigations of the plasma dynamics and ionospheric irregularities using digisondes, with the Centre for Atmospheric Research of the University of Lowell (USA), solar flares with the Observatory of Meudon (France), modelling of the active region atmosphere in the solar corona with the Institute for Astrophysics (Germany), the high-latitude ionosphere with the Geophysical Institute (University of Alaska), the environmental effects on high-orbit spacecraft, together with the ESA, etc.

However, the potential for further development of our scientific and scientific-technical cooperation with scientists from NATO and Western European countries is by no means exhausted. For well-known reasons, it was not until recently that possibilities for extensive cooperation have emerged.

Obvious and promising areas of cooperation may include both the implementation of programs already carried out or submitted to a competition of various foundations, and projects submitted by the ISTP to consideration of the NATO Scientific Committee.

There exists a high potential primarily for pooling efforts in a substantial extention of the monitoring of solar-terrestrial relationships. For this purpose, we have to assimilate and use experience of the Western countries' observatories in regard to recording data on modern data carriers and the generation of data bases, i.e. to modernize recording systems of our instruments and facilities and uniformly organize and generate data bases.

Our Institute is now involved in ground-based tracking of the Western European solar orbiting observatory SOHO [10]. It is possible to significantly promote the research activities at the radio heliographs in Nancay and Westerborgh through coordination of observations there and at the SSRT [11]. Our proposals were sent to the President of the Joint Organization for Solar Observations and the Chairman of WG-6 as well as to the Observatory of Meudon.

We believe that the proposals for pooling efforts in investigations on ecological aspects in the field of solar-terrestrial physics are highly important, including the upper atmospheric ecology, ozonometric monitoring and remote sounding of the upper atmosphere, determination of atmosphere-ionosphere coupling effects, separation of environmental impacts of natural and technogenic factors, etc. We are hoping to arouse interest of the Space Physics Institute (Sweden), the Geophysical Institute of the University of Alaska (USA), El Instituto Nacional de la Tecnica Aerospacial (Spain), Max-Planck Institute of Aeronomy (Germany), the British National Space Centre, the National Centre for Space Research (France), the European Space Agency, the EISCAT Association, and the Environmental Research Laboratories of NOAA (USA).

Based on the implementation of the previous proposals, or in parallel with them, it is appropriate to pool efforts for a joint study of the spacecraft-environment interaction, including contamination of outer space with man-made objects. These topics may be of interest for the Mullard Space Science Laboratory (England), the Belgian Institute of Space Aeronomy, and the European Space Agency.

The ISTP can participate jointly with the medical institutions of Irkutsk in ascertaining mechanisms, by which heliogeophysical and meteorological factors affect the organism of a healthy and a sick man as well as extended airborne and underground engineering lines.

For a particular acquaintance with the Institute, its observatories and specialists, it would be desirable to provide for and realize visits to Irkutsk of experts from the NATO Scientific Committee or from interested agencies, universities and observatories.

Of course, while having attained certain achievements, our Institute encounters significant difficulties of enduring Russia's renewal. We have run into difficulties with maintaining the unique facilities and longterm data archives that are Russia's treasury. In order to use them to advantage, it is necessary, first of all, to re-equip them with modern computing hardware and modernize the recording systems as well as to have means and channels of modern on-line communication for adequate contacts with partners.

References

[1] Rishbeth H., Wills D., Hewish A. (1993). Long-Term Solar-Terrestrial Monitoring. EOS, Transactions American Geophysical Union, v. 74, №28, pp. 313, 317—318.

[2] Sentman D., Troitskaya V., Breus T. (1992). Solar-Terrestrial Effects on Biota. A Plenary Report of the SCOSTEP Ad Hoc Committee on Solar-Terrestrial Effects on Biota. Aug.

[3] A World-Wide Solar-Terrestrial Energy Program by the Ad-Hoc ICSU/SCOSTEP STEP Planning Committee, March 1986.

[4] Alissandrakis C.E., Lubyshev B.I., Smolkov G.Ya. *et al.* (1992). Two-Dimensional Solar Mapping at 5.2 cm With Siberian Solar Radio Telescope. Solar Phys., v. 142, pp. 341—358.

[5] Altyntsev A.T., Grechnev V.V., Zubkova G.N. *et al.* (1983). New Capabilities of SSRT: Observations of Millisecond Microwave Spikes. Abstracts of Int. Workshop "Fragmented Energy Release on the Sun and Stars" (Oct. 18—21, Utrecht), p. 41.

[6] Ledebev V.G. and Urbarz H.W. (1993). The Generation of Solar Type II Radio Bursts in Coronal Magnetic Loops. Solar Phys., v. 146, pp. 365—376.

[7] Kichatinov L.L. and Rudiger G. (1992). Magnetic-Field Advection in Inhomogeneous Turbulence. Astron. and Astrophys., v. 260, pp. 494—498.

[8a] Mishin V., Block L.P., Bazarjapov A. *et al.* (1993). A study of the CDAW-9C Substorms, May 3, 1986, Using MIT-2, and a Substorm Scenario With Two Active Phases. J. Geophys.Res., (in press).

[8b] Mishin V., Opgenoorth H. and Murphrel G.S. (1992). The Methods to Determine the Polar Cap Boundaries. J. of Geomagn. and Geoelectr., v. 44, pp. 1207—1214.

[9] Zvezdin V.N. and Fridman S.V. (1992). Regimes of Ionospheric Turbulence From Fractal Analysis of Satellite Radio Signal Scintillations. J. Atmos. Terr. Phys., v. 54, N. 7—8, pp. 917—952.

[10] Joint Organization for Solar Observations. Annual Report (1992). (Ed. Alvesleben A.V.).

[11] Smolkov G.Ya., Pistolkors A.A., Treskov T.A. *et al.* (1986). The Siberian Solar Radio Telescope: Parameters and Principle of Operation, Objectives and Results of First Observations of Spatio-Temporal Properties of Development of Active Regions and Flares. Astrophys. and Space Sci., v. 119, pp. 1—4.

♦

SPACE CONTAMINATION AND ECOLOGICAL PROBLEMS

V.M. FOMIN, A.M. KHARITONOV, G.A. SAPOZHNIKOV and
E.I. GRINBERG
Siberian Centre of Aerophysical Studies,
Institute of Theoretical and Applied Mechanics of SB RAS

ABSTRACT. The state-of-the-art theories concerning the problem of pollution by space vehicles are discussed and a variety of mechanical problems' formulations are considered, as applied to environmental problems in space.

Keywords: space debris / impact / aerodynamic.

Active international human activity in space produces vast accumulation of various space fragments: spent satellites, rocket stages, engine combustion products, elements of the systems separating satellites from launch vehicles, satellite and rocket fragments resulting from accident or deliberate explosions, and many other types of debris.

Of particular interest are the objects in space found just at the limit of detection for tracking space vehicles or below the limit. Usually, these objects have small dimensions, nevertheless, they can produce a hazardous collision. For example, an aluminum ball of a diameter of 1 cm has the same kinetic energy as a standard grenade. Only a few modern satellites can endure such an impact. By now many narrow misses of various fragments of space vehicles and collisions in orbit have been noted. In 1982 a fragment of a Russian rocket swung near the spaceship "Columbia" at a distance of only 12 km. Such examples are numerous.

Collecting at near-earth orbits, all these fragments make up so-called "space debris". At present about 3.5 million fragments are orbiting the Earth. According to NASA predictions orbits lower than 5500 km will contain up to 6 mln kg of space debris by the year 2010 and judging from the data of the USA National Security Council this figure will be twice as high.

1. Formation and disappearance of space debris

Common sources of space contamination are accidental or planned explosions of launch vehicles and satellites. In 1964-1985 the planned destruction of 34 satellites resulted in the formation of 2 thousand fragments, while the "Ariane" explosion in 1986 resulted in more than 3 thousand catalogued objects [1]. The collision of space debris fragments leads to formation of new fragments. During the last decade the average number of such collisions was two per year [2]. Note that this number is bound to increase due to repeated collisions even if new space vehicles will not be launched at all. It should also be noted that the degree of contamination of

V. A. Koptyug and J. Klerkx (eds.), Science Policy: New Mechanisms for Scientific Collaboration between East and West, 177–184.

the near-earth space varies in different altitudes. The majority of the fragments are located at the altitudes from 800 km up to 1000 km in the region of 1500 km (see Fig. 1) [3].

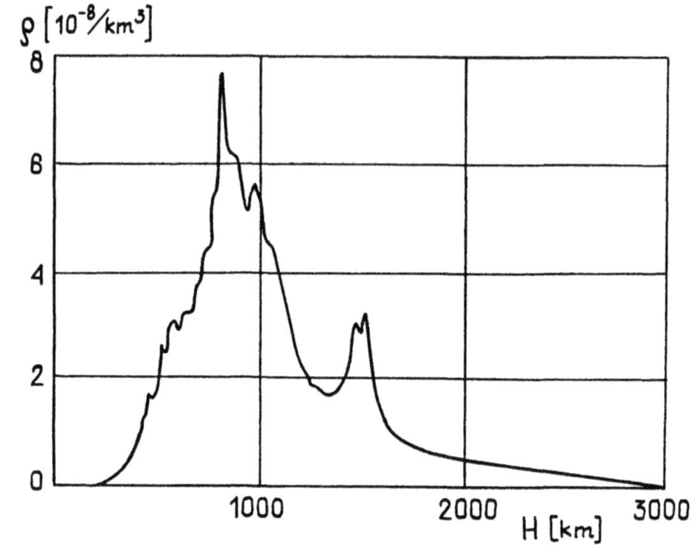

Fig. 1.
Object density as a function of altitude.

Fig. 1 shows the components of the total distribution density of the fragments ρ as a function of altitude H. The fragments also differ greatly in size. The majority of the fragments (99%) are less than 10 cm and 1% — more than 10 cm. These fragments may be detected by modern tracking and, hence, be catalogued. The collision probability is approaching 10%, which requires the detailed study of the consequences of possible collisions. If these consequences prove dangerous for people they must be predicted and corresponding measures must be taken.

There are various approaches for predicting collision processes of space fragments. Along with the satellite orbit parameter one should know the density of space contamination by fragments. So it is necessary to track not only active satellites but also the thousands of fragments which constitute space debris. Therefore, a wide net of ground-based tracking stations is necessary and all trackable space fragments should be catalogued. The compiling of catalogues requires the creation of statistic models of distribution of objects of artificial origin in the near-Earth space. This problem is under study but the capability of technical means is rather limited at the present time [2—6].

Let us consider one of the possible approaches, where all mechanical aspects of the fragments' behavior are included integrally. Let N be the total number of objects located at a fixed orbit. The velocity of the fragments' changes is a number determined by the equation

$$dN/dt = q_1 + q_2 + q_3,$$

where $q_1 = a_1 \cdot exp(-N_1/N)/\tau_1$ is the source of the fragments' number caused by their collisions, $q_i = (-1)^i a_i \cdot (N-N_i)/\tau_i$, $\tau_i = 2, 3$ is the fragments' source or run-off due to space vehicles launch, or the destruction caused by aerodynamic braking in dense atmospheric layers, respectively, τ_i are relaxation times, a_i and N_1 are fixed parameters. N_2, N_3 characterizes the population level at which a catastrophic splitting begins, and are the limiting values of population dimension determining the final equilibrium states. Depending on the choice of the initial data and various scenarios of space activity development, several types of population behavior were classified and criteria of transition between them were given:

— regular (1);
— irregular clearance of space (2);
— catastrophic splitting (3). It is shown in Fig. 2 by dashed lines while the solid lines show the results [7—8] obtained for predicting discrete observations and mathematical simulation using the famous CHAIN program.

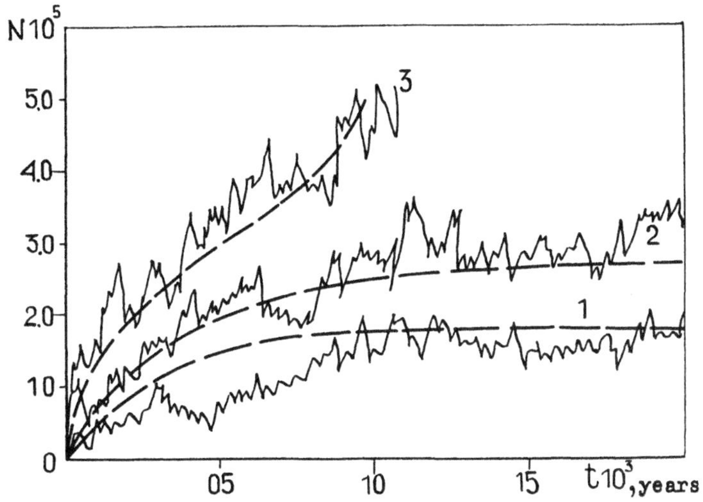

Fig. 2.
Evolution of the population for different increase rates of the basic population.

Therefore, a new principle of space activity should be established: all objects fired into high, near-earth orbits, where no self-clearance effect is observed should be removed at the end of the flight program. The geostationary orbit (altitude 36 km) stands out as an example among others. This orbit is usually used for communication satellites. A space vehicle remains in the same position with respect to the Earth, at this orbit. Space debris hampers the normal functioning of such vehicles at the geostationary orbit, and may exist there for millions of years. Therefore, the contamination of this orbit draws special attention from the International Union on Radio Communication, European Space Agency and NASA [9].

Especially hazardous are the possible collisions of space fragments with nuclear power sources orbiting the Earth. The total number of Soviet space vehicles (SV) with spent nuclear power sources (NPS) is 31. They are transferred from the low orbit to the drift orbit (altitudes of about 800—1000 km). Table 1 shows the parameters of Soviet satellites possessing reactive cores. Beginning with "Cosmos-1176", the reactive core was detached backwards. The duration of the passive existence of satellites and reactive cores is not less than 300 years, and for "Cosmos-1900" not less than 200 years, which is enough for the products of fuel splitting to reach a level of safety. Parameters of American satellites with reactive cores are given in Table 2.

NN	Satellite	Launch date	Drift orbit transfer date	Launch or drift orbit, km	
				perigee	apogee
1	Cosmos-367	03.10.70	03.10.70	915	1040
2	Cosmos-402	01.04.71	01.04.71	945	1032
3	Cosmos-465	25.12.71	05.01.72	930	1027
4	Cosmos-516	21.08.72	25.09.72	910	1036
5	Cosmos-626	27.12.73	10.02.74	912	978
6	Cosmos-651	15.05.74	25.07.74	891	947
7	Cosmos-654	17.05.74	30.07.74	908	1024
8	Cosmos-723	02.04.75	15.05.75	903	960
9	Cosmos-724	07.04.75	21.07.75	854	945
10	Cosmos-785	12.12.75	12.12.75	887	1026

Table. 1.
Orbital parameters of soviet space objects with nuclear power sources.

NN	Satellite	Launch date	Orbit, km		Orbit inclination, deg	Orbit resident time, years
			perigee	apogee		
1	Transit-4A	29.06.61	890	1000	67,5	500
2	Transit-4B	15.11.61	960	1130	32,4	1200
3	Transit-5BN-1	28.09.63	1090	1150	89,9	1900
4	Transit-5BN-2	05.12.63	1080	1110	90,0	1800
5	Nimbus-3	14.04.69	1070	1131	99,9	3600
6	Nimbus-4	08.04.70	1085	1098	99,7	3600
7	Triad	02.09.72	716	863	90,1	137
8	Nimbus-5	11.12.72	1087	1087	99,5	3600
9	Nimbus-6	12.06.75	1099	1112	99,6	3600

Table. 2.
Orbital parameters of american space objects with nuclear power sources.

It is understandable to think that personnel responsible for the development of SV with nuclear power sources thought that the SV resident time at such altitudes was long enough to become harmless for the world community.

However, simple estimates show that if a satellite with a nuclear power source receives a braking impulse, resulting from its collision with a fragment at the altitude 1000 km, its orbit resident time will be reduced to an hour. It should be noted that this problem arose due to intense space exploration, including accidental and deliberate explosion of satellites and launch vehicles within the altitude range 800—1000 km.

2. Principal goals and capabilities of ITAM

To draw reliable conclusions about the ecological safety of the Earth at these orbits, the criteria of the transition between various types of population behavior should be improved, requiring further studies in various mechanical fields. The following problems should be mentioned first:

1. Prediction of space debris population and estimation of the probability of collisions with space bodies for specified flight conditions;
2. Simulation of the process of fragments collision with space vehicles or their elements;
3. Identification of the secondary field of fragments and determination of their aerodynamic characteristics in free-molecular, transitional and continuum flow regimes;
4. Computation of trajectory parameters of the flight and descent of an object from any orbit, calculation of the time of existence before entering the atmosphere;
5. Prediction of the particles deposition onto the Earth;
6. Search for rational ways of destroying dangerous space debris.

The wide range of several parameters at SV collisions hampers the current creation of a simple theory to reliably calculate the degree of fragments' destruction, the range of sizes and shapes, direction and velocity of flight, the impulse loss value after collision and, consequently, their orbital departure and further behavior.

However, it should be noted that the Institute of Theoretical and Applied Mechanics, the Siberian Branch of the Russian Academy of Sciences, and the International Center of Aerophysical Studies, have highly skilled personnel, scientific potential and corresponding experimental facilities:

- wind tunnel complex (the Mach number range 0.3—25, the Reynolds number range $10^4 - 10^8$, electrodynamic and gasdynamic accelerators with a set of measurement complexes (Figs. 3—4);
- software for numerical simulation: of conditions in free-molecular and continuum flow regimes; of high-impact collision processes; of trajectory parameters of objects' flight and descent from the orbit and the time of their existence before entering the Earth's atmosphere.

Let us consider some examples which reveal the capabilities of the Institute's scientists and simultaneously answer some mechanical problems in near space ecology.

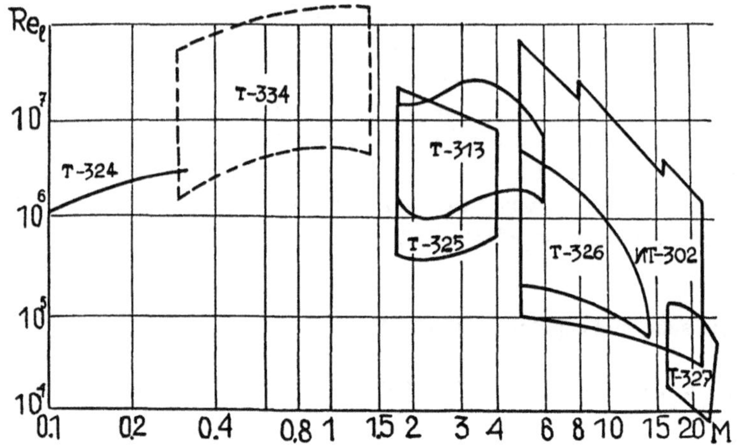

Fig. 3.
The modelling conditions in wind tunnels of ITAM.

Code	Type	Test section size, (m)	Mach range	Stagnation pressure, atm	Stagnation temp., K	Max. Reynolds number/	Running time	Reference
T-324	closed circuit	1 x 1	0.03 ÷0.3	≈	300	3×10^6	60 min	low turbulent
T-334	closed circuit	0.62 x 0.62	0.3÷ 1.2	up to 9	80-300	3.5×10^8		under construction
T-313	blow-down	0.6 x 0.6	1.8÷ 6.0	12	up to 700	6×10^7	5 min	
T-325		0.2 x 0.2	0.5 ÷4	12	300	2×10^7	10 min	
T-326		0.2	5÷ 14	120	up to 1500	4×10^6	1 min	plasma heating
T-327	with vacuum capacity	0.22	16÷ 25	200	2500	2×10^5	40 s	nitrogen
T-333	blow-down	0.3	2÷5			5×10^7	10 min	
UT-302	hot-shot	0.3	5÷ 15	1500	3000	$\approx 10^8$	120 ms	

Fig. 4. Main characteristics of wind tunnels ITAM.

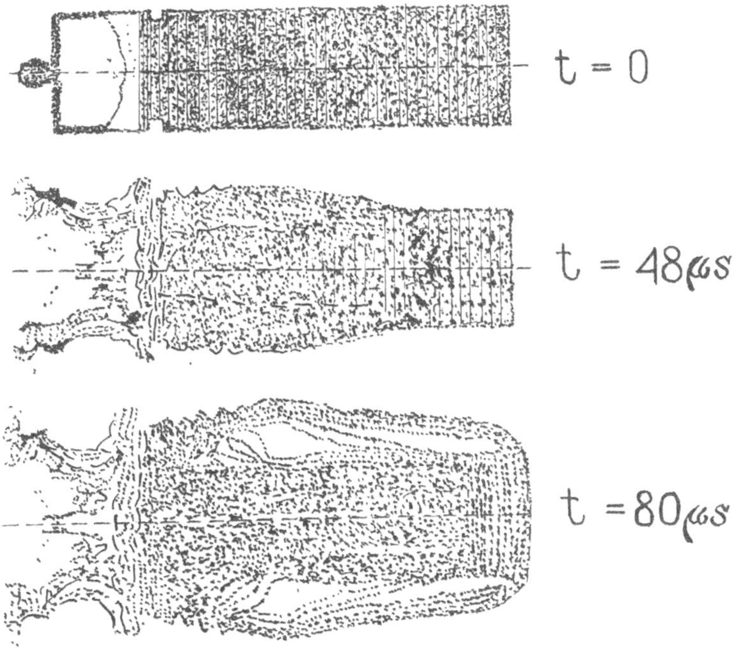

Fig. 5.
Typical pictures of hypervelocity impact.

Figure 5 shows the deformation and probable destruction of the reactive core in the case of a central shock by a spherical fragment m = 90 g and V = 7.8 km/s at various moments in time. The field of fragments formed by an SV collision with various space clouds is often hard to identify. But fragments with various geometric shapes share overall aerodynamic characteristics with the known, well-studied reference bodies. Figure 6 gives such a comparison for cone bodies with a different base for drag ($C_{x\alpha}$) and lift ($C_{y\alpha}$) coefficients, obtained from the experimental data in the wind tunnel T-313 for $M=4$. Similar results are obtained for pyramids. These results considerably simplify further trajectory computations.

The Institute of Theoretical and Applied Mechanics, teamed with *NPO "Krasnaya Zvezda"* (*Red Star*), have carried out a large volume of such investigations. The experimental data obtained allowed us, together with the Gettingen department of DLR, to make up a computer version of an aerodynamic database for various space objects models. So, the results show the danger of space contamination for the nature, and the necessity to clear the Earth's environment of space debris, as well as the importance of the reasonable coordination of various countries in conquering and using space with the minimum amount of harm for nature.

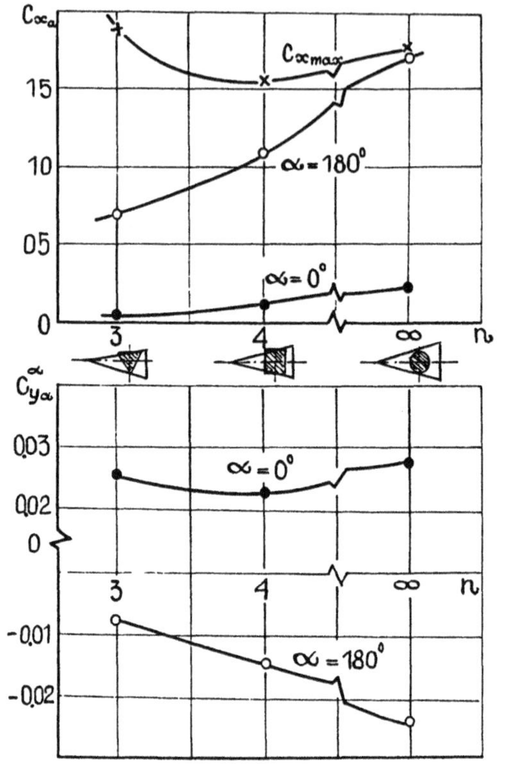

Fig. 6.
Aerodynamic characteristics for some space debris configurations.

References

[1] Grinberg E.I. (1992). Space Contamination and Space Flights. Priroda ("Nature"), № 8.

[2] Lee J., Buden D., Angele J. *et al.* (1990). American Institute of Aeronautics and Astronautics Paper, № 1368.

[3] Rex D., Zhang J., Eichler F. (1990). A Review of Orbital Modeling in Europe. AIAA Pap., № 1354.

[4] Ojakangas G.W. (1990). An Analytic Description of Debris Cloud Evolution Starting from the Equations of Motion. AIAA Pap., № 3904.

[5] Ojakangas G.W., Auz-Meador P., Reynolds R. (1990). Orbital Debris Environment. AIAA Pap., № 3863.

[6] McKnight D. (1992). Debris Growth Isolating Sources and Effects. AIAA Pap., № 1284.

[7] Eichler P., Rex D. (1990). Debris Chain Reactions. AIAA Pap., № 1365.

[8] Eichler P., Bade A. (1990). Removal of Debris from Orbit. AIAA Pap., № 1366.

♦

COOPERATION IN ENVIRONMENTAL FLOW PROBLEMS OF AEROSPACE SCIENCE

G. E.A. MEIER
DLR, Institut fur Stromungsmechanik,
Göttingen, Germany

V.M. FOMIN, A.M. KHARITONOV
Siberian Centre of Aerophysical Studies,
Novosibirsk, Russia

ABSTRACT. Some of the global environmental problems result from the air traffic and from manned or unmanned space missions. Mainly pollution of the atmosphere, noise problems and safety problems have to be named here. A lot of these problems are connected with fluid dynamics which science also provides solutions of the problems or at least improvement of the present conditions. - Since all these problems are of global character and since in the eastem countries a good background in aerospace science is available and the whole field is not highly competitive, this area is especially well-suited for cooperation between east and west.

Keywords : space debris / drag reduction / pollution of atmosphere / noise problems / safety problems / fluid dynamics / aerospace science.

1. Introduction

The cooperation in aerospace sciences has for the fluid dynamicist namely two interesting aspects.

- environmental research in the field of pollution and noise
- advanced technologies of propulsion and aerodynamics to improve the performance of aerospace vehicles.

In these two fields aerodynamics play an important role in connection with a lot of other disciplines of aerospace science. So the cooperation between east and west in this fields gives also the opportunity to include contributions of different research teams which are not all exactly aligned on the same topic.

Since for good and well organized cooperation the exchange of data and results and common experiments and projects are very important, the contracts for the future programs should include especially resources for conferences, visits and communication.

2. Atmosphere and Space Ecology

This field is mainly devoted to the observation and analysis of the environmental pollution and the safety problems connected with space debris which is all kind of man-made waste in orbit or descent.

V. A. Koptyug and J. Klerkx (eds.), Science Policy: New Mechanisms for Scientific Collaboration between East and West, 185–192.

2.1 SPACE DEBRIS

The number of operational satellites and defunct relicts in orbit approaches 10000 even for the size trackable by camera or radar, and grows by orders of magnitude when counting into the sub-mm size range. Therefore space debris has been recognized as potential source of trouble to space missions and on ground, for reasons explained in the following.

2.1.1 Problem Description. With time in orbit, debris tends to disperse into expanding bands of altitude and inclination, influenced by various perturbations (earth and moon gravitation, atmospheric drag, radiation pressure) and depending on its origin. Therefore collisions with spacecraft may occur at high relative speeds (> 10 km/s) and present, especially for manned missions, an unacceptable risc of catastrophic damage due to extreme impact energies. For the U.S. Space Station e.g., the probability of such collisions has been rated high enough to modify original plannings.

Some objects are massive enough not to burn up completely when their orbits have shrunk to final decay by atmospheric drag (e.g. Skylab and Salyut-7), such presenting a potential hazard to populated regions on impact.

Others carry nuclear power sources, which may severely contaminate atmosphere and ground by radioactive matter when melting up during re-entry (e.g. Kosmos 1403).

2.1.2 Space Debris Monitoring. Permanent monitoring of space debris is difficult due to the sheer number of objects and to the size limits of optical and radar tracking. Expecially the smaller of them continuously change orbits under gravitation, atmospheric drag and radiation pressure nearly unpredictably, as their material and surface properties are unknown and their attitude is uncontrollable and mostly unknown, too.

Cases have already been recorded of mutual collisions between debris objects, generating showers of secondary debris with distributions of fragment size and orbit parameters that can be estimated statistically at best.

With increasing density of debris, even some kind of "collisional chain reaction" may occur, completely contaminating certain orbits and making them unusable.

Limited or lacking information on object properties and lacking controllability also hinder tracking and prediction of the final decay of debris due to atmospheric drag. Whether disintegration and bum up by aerothermal loads or impact in one or more fragments occur, how nuclear power sources melt down and release their radioactive contents contaminating the atmosphere, and where the ground is finally hit, can hardly be determined in advance with the accuracy satisfying safety requirements.

2.1.3 Gaining Control of Space Debris. The attached non-exhaustive list of (somewhat randomly) selected references gives a hint to the numerous national and international research activities on space debris, not to mention political ones to develop future prevention strategies.

International cooperation is essential to provide the necessary exchange of information on constructive and operational parameters of debris objects, and the global tracking capabilities to reduce any danger to space missions and to earth environment.

With the sound knowledge in the problem area already available in Europe and in Russia, and the long active experience in all kinds of space missions from the latter, a collaboration of specialists from fluid dynamics, material science, atmospheric science etc. to treat all aspects of the space debris problem promises to be fruitful and successful.

2.2 ATMOSPHERIC POLLUTION BY PROPULSION SYSTEMS

The atmospheric pollution by traffic and industry is one of the most severe environmental problems of the present time. Flows have especially great influence on the propagation, the distribution and the emission of pollutants.

2.2.1 Aircraft Pollution Problems. The modern aircraft has reduced the consumption of fuel already dramatically but there is potential of another reduction of about 50% in fuel consumption by different means concerning the combustion process and the efficiency of engines and also the aerodynamic drag of the whole aircraft. Beside this very important area of cooperative work also the distribution of pollutants in the earth atmosphere is a good topic for international observation and considerations about a more tolerant distribution of pollutants in the atmosphere. In the neighbourhood of airports research on the propagation of exhaust plumes and the emission of fuel vapours is another important topic.

2.2.2 Atmospheric Pollution by Spacecraft. Although the number of space vehicles is compared to the number of cruising aircraft rather small it is of considerable interest, how much pollution is released by rockets and other space vehicles to the atmosphere. Especially the difference between solid propellant rockets and the other propellant systems operating with liquid propellants is of interest in connection with a comparison of the impact on the atmosphere.

2.2.3 The Pollution by Earthbound Vehicles. Earthbound vehicles like cars, trucks, trains and ships have to be considered as mobile pollutant generating sources, the exhaust of which is distributed by wind and motion of the vehicles in widespread areas of the atmosphere before they influence settlements, woods or fields. Very often these propagation mechanisms do not stop at the country's border so that complicated international conflicts are resulting from that. Here a joint cooperative research would be needed to provide the governments with fair investigation results and judgements.

2.2.4 Pollution by Industry and Heating Devices. Industrial plants, the heating of buildings and also natural fires produce most of the carbon dioxide which mainly increases the heating of the earth atmosphere. International cooperation could provide monitoring systems and also ideas for better use and distribution of energy resources.

2.3 MULTIDISCIPLINARY EARTH AND SPACE MONITORING BY SATELLITES

The numerous satellites in different orbits allow for a very accurate observation of all environmental and geological events and properties in all parts of the world. Especially the eastern countries, first of all Russia, have a lot of monitoring satellites which could be used in common evaluation and observation programs. The recorded data could be used in many other common research projects in different fields.

Another use of communication satellites as presently for most of them considered would be an active role in communication and data exchange between the cooperative research projects between east and west. Especially the nowadays still fairly poor terrestrian communication links could be improved. So one of the major drawbacks of present cooperation could be eliminated.

2.4 AIRBORNE AND TRAFFIC NOISE

The noise problem is increasing with industrialization and growing traffic. In this field international standards and rules as well as measures to reduce the noise production of existing machinery are highly appreciated by mankind. Especially emission and imission research in the field of sound and noise is needed because a lot of the problems can be solved by simple measures of reducing the sound generation, by screening and by optimization of vehicle passes.

2.4.1 Jet Noise, Airframe Noise and Sonic Boom. The most pronounced sound sources are beside explosion and other occasional accidents the relatively unscreened sound sources of aircraft. The severe problems of jet noise and propeller noise have been reduced a lot by modem high by-pass ratio turbo engines. Here the low velocity of the jet is favourable for an effective noise reduction. Although the achieved progress is remarkable there is still a lot to do, especially if a supersonic transport is considered. Here the problem of sonic boom has to be solved, otherwise an operation of such an aircraft will not be possible. The helicopter noise is another field, where the impact on the population of cities is very annoying. This problem has also to be solved in the near future by intemational cooperation, because the helicopter as a rescue vehicle will be of increasing importance in health and traffic service. The noise of vehicles like cars, trucks and trains is a severe problem for neighbours of the main traffic routes. Under favourable atmospheric conditions the traffic noise can be propagated over large distances. Here once again the research in these emission problems is very much needed because the increasing traffic predictably results in increasing noise.

2.5 RENEWABLE AND PERSISTING ENERGY RESOURCES

A lot of the previously described problems can be solved or decreased in importance if the quality of energy resources is increased. This means that the use of renewable or pemmanently persisting resources like water, wind, geothermal and solar energy in increasing quantities will help to solve the environmental problems also. Especially in cooperation with the eastem countries, where wide areas with low population exist the wind and water energy sources could be further developed. Space for big power-plants of this kind would be available there. This would be also especially important with respect to the substitution of old atomic power-plants which result in dangerous conditions not only for the country or area.

3. Development and Application of Aerospace Techniques

The whole field of advanced technologies is very interesting for cooperation between east and west because of the long lasting separated independent development of technological ideas and procedures. There is a lot of techniques which can be further developed in bilateral or multilateral projects. In this article some flow related techniques will be addressed which mainly contribute to aerospace science and industrial programs.

3.1 DRAG MECHANISMS AND DRAG REDUCTION FOR VEHICLES

The aerodynamic drag of transport vehicles is one of the big energy sinks which cause at first fuel consumption and at second the before-mentioned pollution problems. Therefore drag reduction means are one of the major technological goals in all aerodynamic research programs. But because the matter is scientifically extremely difficult and a lot of different ways for solutions seems to be possible, an international cooperation is very useful to increase the intensity and the variety of research programs.

3.1.1 The Different Aerodynamic Drag Mechanisms. All vehicles moving in a fluid develop boundary layers and wakes which have a strong influence on the pressure distribution of the surrounding flow field. From this pressure distribution results the so-called pressure drag which is the main reason for the drag of blunt vehicles. But there is also friction in the boundary layers which contributes to the drag by shear forces. Especially the latter one depends a lot on the laminar or turbulent state of the boundary-layer flow. The mechanisms of transition to turbulence for all kind of flows, independent if subsonic or supersonic flows, are not very well understood and are an area of international cooperative research. Also the mostly unsteady separation of boundary layers and the above mentioned formation of vortical flows in wakes are not understood completely. The unsteadiness of wake flows results additionally in vibrations of the vehicles and safety problems for the following traffic. The understanding of these mechanisms and the ways of possibly influencing them are the tools for all technical improvements in the field of drag reduction.

3.1.2 Drag Reduction Devices. By the ongoing research in east and west some drag reduction devices have been developed already. The research on laminar airfoil sections, laminarization techniques by suction of the boundary layer and by riblets, as well as new methods to reduce the strength of shock waves in supersonic flows, can be increased by common research activities. All these measures are of high economical and ecological importance.

3.2 SPACE VEHICLES AND SATELLITE AERODYNAMICS

The aerodynamics in this field of supersonic and hypersonic flight in rarefied atmosphere is a special field which has got increasing importance with the increasing activities in space and orbit.

3.2.1 Aerodynamics of Space Vehicles. There is still a technological competition between rockets and airbreathing space vehicles in the flight through the lower atmosphere. Because of the tremendous cost of research and development in this field the cooperation between the well-established space research agencies in east and west is highly recommended. Also in this area the main problem is, how to lift a pay-load with a minimum amount of energy into earth orbits or into space. In case of airbreathing vehicles the hypersonic flight aerodynamics and the technology of airbreathing engines, which have to operate over a wide range of cruising velocities, have to be developed. In case of rockets some important questions of the outer aerodynamics, the combustion, and nozzles have to be solved also. Problems of safety and re-entry are existing in both cases and can profit a lot of the combined international experience in east and west.

3.2.2 Satellites Aerodynamics. The crude assumption that satellites are moving frictionless in the vacuum of space is not true at all, because most of the orbits at least partially come into contact with the more dense atmosphere at altitudes down to about 200 km. There are mainly three problems which have to be considered in the future.

- The first one is the aerodynamic drag problem which reduces the lifetime of satellites because drag results in loss of height, which has to be compensated for by additional power of the thrusting rockets. Therefore the lifetime of low orbiting satellites is dramatically shortened by aerodynamic drag.
- Another problem is related to the attitude stability of satellites, because the huge solar-cell paddles of satellites form effective torque generating devices which can also be used intelligently for steering purposes.
- A third problem are surface contamination and surface reactions on satellites by the interaction with the impinging molecules. Because of the high relative velocities the gas-surface interaction is a very delicate problem which is not understood up to now.

3.3 AEROTHERMODYNAMICS OF HYPERSONIC TRANSPORT

Closely related to the foregoing chapter the aerothermodynamics of the flow and the boundary layers of re-entry space vehicles or also of supersonic transport vehicles is one of the limiting factors in technical design.

3.3.1 Shock Structures and Wave Drag. The shock structure of flows of supersonic and hypersonic vehicles is of major influence on drag and heat generation by the flow. The shock waves which surround the front of the whole vehicle depend in shape and strength of the geometry of the vehicles body. By special shapes for instance the shock strength can be optimized for minimum drag at constant weight or pay-load of the vehicle. Also the heat production by the shock-wave system, which is determining the heat-load on the different portions of the vehicles structure can be influenced by the shape dramatically. The search for aerodynamically and, with respect to heat production, optimized structures is not finished up to now. At present once again wave-rider concepts are considered as an optimal solution. Since there is a lot of experience from former considerations in this field, especially in the east, a cooperation would be very useful.

3.3.2 Boundary Layers of Hypersonic Vehicles. The boundary layers of super- and hypersonic vehicles tend to be laminar in the first part of the body because of the typically low Reynolds numbers even for huge vehicles. Therefore, the problem of transitional boundary layers and the influence of the boundary layer status on heat production and cooling is very important. The cooling problem for descending space vehicles and for supersonic transport vehicles in the denser lower atmosphere is a crucial problem because for instance up to 99% of the kinetic energy of a space vehicle have to be transfered to the surrounding flow by heat transfer or ablation cooling. Similar problems exist also for supersonic aircraft.

3.4 HEAT TRANSFER AND MATERIAL PROBLEMS

The way of heat transfer is not only, as mentioned above, an outer problem, with respect to the transfer to the surrounding fluid, but also a material and structural problem for the remaining heat, which has to be transfered to the inner structure to prevent the surface from melting or burning.

3.4.1 Heat Transfer from Space Vehicles. The influence of surface properties on heat transfer to the surrounding flow is mainly determined by the radiation properties at higher temperatures and the surface roughness, because this is influencing the degree of turbulence in the flow. This means, that despite of the heat conductivity of the material the radiation cooling is very important as well as the laminar or turbulent state of the boundary layer.

3.4.2 Material Structural Problems. The surface material determines, as mentioned above, the heat radiation properties of the surface into the surrounding flow. In this area a lot of research has been performed already in the space research centres in the east. But also the material and structural properties of the surface have been considered there up to special tiles which have been used in several space or re-entry missions. These questions, which are highly important for the heat insulation and temperature distribution on- and inside of the surface, have to be treated more elaborately in the future. For a capsule technique of re-entry especially the problem of ablation cooling and special questions of optimal shape have to be treated for future space activities.

3.5 FUEL SAVING PROPULSION TECHNIQUES

For all kind of transport but also for industrial heating the most efficient and pollution-free combustion techniques are still a high ranking task because of environmental problems. The combustion of gas, fuel and coal will remain one of the main energy sources for the next future and therefore the improvement of efficiency and cleanness of combustion and the most efficient use of the produced mechanical energy are very important goals of the international research.

3.5.1 Combustion in Engines and Heaters. In the field of combustion the increase of efficiency is rather limited because there are only very little margins left. But better evaporation of fuel and its mixing with the air for combustion can result in further but small increases of efficiency. But these techniques, together with an accurate control of the chemical reaction, will also be used to avoid the generation of the major disturbent pollutants like nitric oxides. The production of carbon dioxide can be reduced by conversion to fuel with less carbon content like for instance the hydrogen technology.

3.5.2 Aerospace Propulsion Techniques. The propulsion techniques for space vehicles are mainly rockets and possibly in future airbreathing engines for the atmospheric flight period up to the space. Also the rockets have to be improved with respect to efficiency and generation of polluting substances. But for the case of airbreathing jet engines a lot of problems connected with the different operational conditions have to be solved. The wide range of air densities and Mach numbers during flight have to be considered in very delicate technical solutions. In the field of airplanes, especially for the new generations of very large aircraft, the efficiency of propulsion has to be increased by higher by-pass ratios and counter-rotating fans of the turbo engines. In this very costly and ambitious area of technological development some common research programs between east and west are already on the way and the hope is that this and a lot more of other cooperative work can be increased or established in the future.

4. References

1. Boettcher, R.-D.: Space environment. DLR-IB 91 A 24, Gottingen 1991.
2. Eichler, P.; Rex, D.: The risk of collision between manned space vehicles and orbital debris - Analysis and basic conclusions. ZfW, Vol. 14, No. 3, 1990.
3. Potter, A. et al.: Summary of the AIAA/NASA/DoD orbital debris conference: Technical issues and future directions. Advances in space research, Vol. 11, No. 12, 1991.
4. Schumann, U.: On the effect from aircraft engines on the state of the atmosphere. AGARD, 81st Symp. on 'Fuels and Combustion Technology for Advanced Aircraft Engines' in Colleferro near Roma, Italy, 10-14 May 1993.
5. Schumann, U. (Ed.): Air traffic and the environment - background, tendencies and potential global atmospheric effects. Proc. of a DLR Intern. Coll., Bonn, Nov. 15/16, 1990, Lecture Notes in Engrg., Vol.60, Springer-Verlag, Berlin.
6. Shine, K.P. and Sinha, A.: Sensitivity of the earth's climate to height-dependent changes in the water vapour mixing ration. Nature 354, (1991).
7. Weyer, H.B.: Technologie fur kunftige schadstoffarme Luftfahrtgasturbinen. Deutsche Physikalische Gesellschaft, Conf. Proc., 1992.
8. Space debris: a potential threat to space station and shuttle. United States general accounting office report, GAO/IMTEC-90-18, Washington, D.C., April 1990.
9. Safety aspects of Nuclear Power Sources (NPS) for electricity generation in space. Report FuE-87062, NUKEM GmbH, Hanau and Institut fur Reaktor- und Raumflugtechnik TU Braunschweig, Feb. 1988.
10. Proceedings of the workshop on space debris. May 11th, 1989 at ISAS. The Institute of Space and Astronautical Science report SP-ll, Kanagawa, Japan, March 1990.
11. Space Debris - The report of the ESA space debris working group, ESA SP-l 109, Nov. 1988.

5. Acknowledgement

The authors have to thank R.-D. Boettcher for valuable assistance.

LONGTERM COOPERATION OF SB RAS AND BRUKER-SPECTROSPIN GROUP OF COMPANIES: RESULTS AND PROSPECTS

R.Z. SAGDEEV and A.V. PODOPLELOV
Siberian Tomography Centre, Novosibirsk, Russia

G. LAUKIEN
Brüker Medizintechnic, Karlsruhe, Germany

ABSTRACT. The development of relations between Bruker-Spectrospin Group of Companies and SB RAS, which resulted in the foundation, construction, and equipping of a new research institution of SB RAS - the International Tomography Centre, is described. Problems of cooperation of the Academy of Sciences and a private firm are discussed.

Keywords: international cooperation / medicine / inspection of population.

The German firm Bruker (or, more exactly, Bruker-Spectrospin Group of Companies from Germany, Switzerland, and France) is known as one of the world leaders in the production of analytical and scientific equipment. The Siberian Branch of the Russian Academy of Sciences is a long-standing partner of this firm. The latter has supplied the institutes of SB RAS with a large variety of instruments: NMR and ESR spectrometers, chromatographs, IR-spectrometers, and others.

In many cases the equipment was delivered and installed before payment. In addition, the firm has held in the SB RAS a number of demonstrations, exhibitions, and workshops.

The next leap forward in the cooperation of Bruker with the Siberian Branch was the installation of a medical NMR-tomography system in 1988 (with payment by installments). It was the first system of its kind in Siberia. In the same year the system was put into operation and diagnostic examination of patients from all.

Siberian Regions began.

At the same time the idea was conceived of the cooperation of the new type: establishment of a joint international institute (International Tomography Centre) as a scientific research. This idea was very timely as it coincided with our "perestroika" and with changes in financing the academic science in our country.

To realize this idea, the Institute of Chemical Kinetics and Combustion together with the Presidium of the Siberian Branch of RAS set up the Board of Construction of International Tomography Centre, which took up the functions of general contractor and sponsor of the construction. From the very beginning, Bruker-pectrospin Group of Companies contributed

V A Koptyug and J Klerkx (eds), Science Policy New Mechanisms for Scientific Collaboration between East and West, 193–196

SF 1172000 to the construction of the building on account of the future authorized capital. By 1993 the construction of the building and its equipment (with instruments, devices, etc.) had been finished and the International Tomography Centre was registered officially as a separate institute. ITC includes six scientific laboratories conducting fundamental research in various fields of chemical physics, an engineering department, and a department of medical diagnostics. The scientific laboratories of ITC carry out fundamental investigations in the following fields: magnetic spin effects in chemical reactions of radicals; design of new magnetic materials, organic ferromagnets; development of new contrasting agents for NMR-tomography; theoretical problems of chemical physics. Aside from scientific work, an integral part of ITC activity is the diagnostic examination of patients on the NMR-tomography System. Among the important tasks facing ITC is the assistance in the development of this method in Russia and in other countries of the Commonwealth of Independent States. Figures 1, 2, 3 show the general view of ITC building, examining of a patient by Magnetic Resonance Imaging (MRI) method, and a clinical case.

Fig. 1.
New building of ITC for fundamental researches and administration.

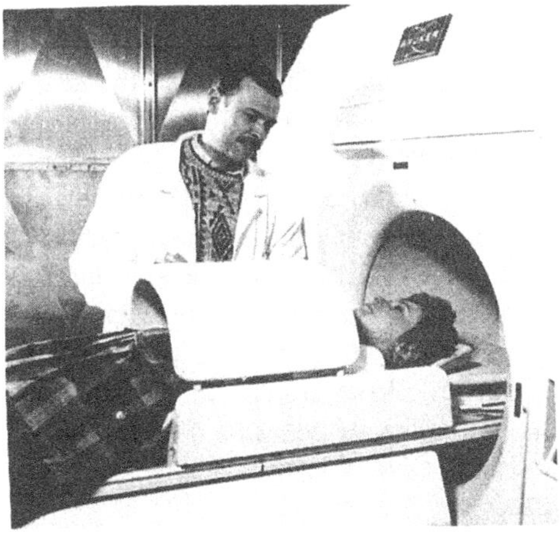

Fig. 2.
MRI-examinating performed in the medical diagnostic department.

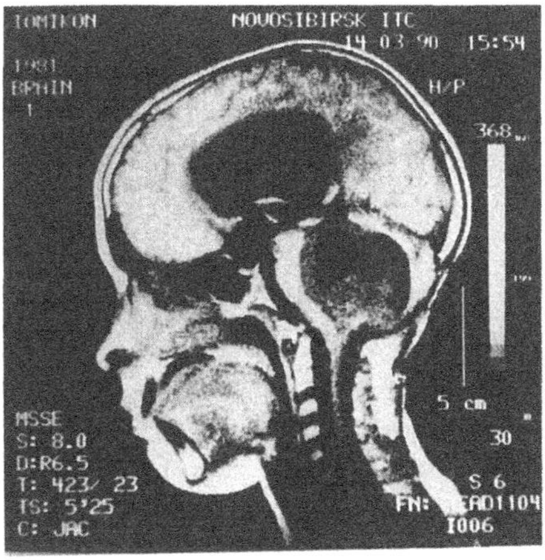

Fig. 3.
Tumor Brainstem, sagittal plane.

ITC has established scientific contacts with a number of internaional scientific centres and participates in joint scientific projects.

Financial resources of ITC include the following: grants for special scientific projects, applied work earnings, incomes from maintenance of medical and analytical equipment as well as from training of personnel and marketing. Besides, the total contribution from Brüker is 2 million 371 thousand SF (supply will be completed by September 1994). The Russian side contributed the equal sum (in currency).

In conclusion, it should be noted that the tasks of the private firm interested in getting profits are in conflict with the problems facing the Russian Academy of Sciences whose primary interests are in the development of fundamental research which effects profits not as quickly. Therefore, the most difficult point in such cooperation is balancing the interests of both sides. According to the Charter of ITC, this institute is a profitless organization performing fundamental and applied studies. However, the Brüker firm gets indirect profits due to the advertising, marketing, and servicing activities of ITC. In addition, the groundwork has been laid for producing medical and analytical equipment.

♦

II.4. SOCIO-ECONOMIC AND HUMANITARIAN STUDIES

INTERNATIONAL COOPERATION IN STUDIES OF ECONOMIC DEVELOPMENT OF SIBERIA

V.E. SELIVERSTOV
Siberian International Centre for Regional Studies, Novosibirsk, Russia

S.A. SAVEY
Université Paul Valery, Montpellier, France

ABSTRACT. The article describes main directions of cooperation between the Siberian Branch of the Russian Academy of Sciences (SB RAS) and foreign research centres and universities in studying problems of economic development of Siberia and retraining executive and legislative personnel from Siberian regions. This cooperation takes into account developmental tendencies and prospects of Siberian economy and also new realities of the economic and political life in Russia. Information is provided about the activities of the Siberian International Centre for Regional Studies, and concrete research and training projects specific for Siberian problems, which have been awarded foreign grants and are carried out in collaboration with universities of the USA, the Netherlands and France.

Keywords: Siberia / economic and political reform / regional policy / territorial development / scientific cooperation / training of personal.

At the present time the results of the reform of political and economic system in Russia depend to a great extent on the successful realization of real federalism, effective regional policy, the arrangement of interregional relations, the development of the market mechanism and local government institutions in the various regions. These results are largely determined by dynamics and efficiency in the development of such a region as Siberia being unique in the world on the scale and variety of its natural resources. Siberia is also the most attractive region in Russia for international economic and scientific cooperation.

This paper concerns the questions of international complex studies on the socioeconomic problems of the development of Siberia organized by the Siberian International Centre for Regional Studies (SICRS) and the Institute of Economics and Industrial Engineering (IEIE), Siberian Branch of the Russian Academy of Sciences (SB RAS).

V A Koptyug and J Klerkx (eds), Science Policy New Mechanisms for Scientific Collaboration between East and West, 199–208

1. The principal problems of the contemporary Siberian economy and the systems of social-economic studies in SB RAS

At the Siberian Branch of RAS the social and economic studies of the development of Siberian regions are for the most part concentrated in the IEIE of SB RAS and its regional branches. These studies compose the social and economic part of the complex scientific and technical program "Siberia". They rest upon a broad experience of studies, a highly skilled personnel (about 150 doctors and candidates of science are involved in the work in this direction), modern research tools (including original complexes of economic and mathematical models, computer informational and analytical systems).

In recent years Siberian economists and sociologists have gained significant scientific results which characterize feasible strategies for the development of Siberia as an integral part of the country's economic complex, and indicate the main directions of the improvement of economic mechanism. The principles and models for locating new enterprises in Siberia have been offered as well as the scientific grounds for the development of the main fuel and energy base of the USSR, West Siberia oil and gas complex namely, and bringing to the exploitation of Baikal-Amur main line zone and other important territorial-production complexes of Siberia.

These studies combine both the elaboration of new fundamental theoretical problems in the field of regional economy and the preparation of concrete proposals for federal and local authorities and Siberian enterprises. In this respect the economists and sociologists of Siberia have always had their own independent scientific position. Here we should note their efforts have demonstrated the inefficiency of the project to turn the flow of some Siberian rivers into Kasakhstan and Middle Asia. Although the conclusions, which were made on the basis of precise economic and mathematical calculations and on the elaborations of alternative projects to solve the problems of water supply of the southern republics of the USSR, went against the viewpoint of the former USSR administration, the project was stopped due to the principle position of the Siberian Branch of RAS.

Scientific elaborations of social and economic problems on the development of Siberia were the stem of All-Union conferences, held once every five years, on the development of the productive forces of Siberia. The latest conference (now All-Russian) took place in summer 1993, for which IEIE of SB RAS had prepared a comprehensive report reflecting the modern vision of the major problems of development of Siberian regions under the reform of economic and political system in Russia.

Surely, the specific character and novelty of the present situation in Russia have required the change in orientation of the system of social and economic studies in SB RAS. The forecasts of the long-term development of industrial branches and the regions of Siberia, the study of the development and interrelationships of Siberia within the common national economic complex have markedly given way to the problems of the economic reform in Siberia and to the creation of a new economic mechanism considering the actual situation of the contemporary economic and political life in Russia. Therewith, it is of great importance to establish new tendencies in the development of Siberian economy, the ones which are radically different if compared with those of other large regions of this country. These tendencies may involve the following:

Firstly, it is the relationship between Moscow and Siberia and the manner of state regulation which substantially influenced Siberian economic development. The state retained its leading role amidst conditions of collapse of the centralized control over the economy and the chaos in price determination, it brought about the control over production, the degree to which financial resources were withdrawn and the restriction on the behaviour of the producers in Siberia to be kept on a higher level than in other regions of the republic.

At the same time the advantages, in the first place the priority in distribution of the investments and logistics, have practically disappeared. The rapid decrease in the centralized state investment in the basic branches of the Siberian economy has occurred, adversely affecting not only Siberian but also the Russian national economy. The main results of this stage is that the economy has entered the phase of reduced production; i.e. "the living" on accumulated means of production has begun.

Secondly, in recent years a considerable differentiation of the regions of Siberia according to the level of development and the standard of people's well-being have occurred.

Thirdly, the specific character of the moment involves the emergence of a new geopolitical situation in Siberia, which has caused new transport, economic and political problems in the regions. Formerly, when discussing the riches of Siberia which are necessary to be put into circulation, we fairly often neglected the fact that the vast territory of Siberia was its own kind of wealth: to wit, Siberia could serve as a unique bridge connecting West Europe with the states of South-East Asia.

Nowadays the task of the economists lies not only in establishing the tendencies and estimating how acute the new situation for the region is when the state has practically left Siberia to the mercy of fate, but also to discover the strong points where to help Siberia find a way out from the crisis and place it on a trajectory of sustainable development.

We believe that the main remedies to cure Siberian problems are the following:

— new regional policy of the government, in which the state protectionism regarding Siberia sees it as a region, on which the destiny of Russia really depends, should be a nucleus;
— new policy in the economy of mineral resources utilization (a transition to market-based principles of the mining and exploitation);
— a considerable intensification of the economic integration of Siberian regions;
— new structural-investment policy. In particular, to accelerate the investment (including the involvement of foreign capital), it is necessary to use the system of the mortgage. In this process the governmental interference, regulation and legitimacy namely, is badly needed;
— new social policy which will solve the problem of the strong disparity and the inequality in the incomes and the living standards between the different territories of Siberia and various populations. In Siberia these disproportions have achieved 10-fold difference which may not be explained by economic, social and moreover by moral expediency. It erects not only new "distance barriers", which eventually put the population of many, especially provincial regions, in a discrimination position, but sets up the barriers between the regions, their population, between the miners and industrial engineers, oil workers and teachers. It, surely, does not agree with new integration connections between the republics, territories and regions of Siberia;
— new ecological policy involves the fulfillment of the principles and mechanisms of sustainable development which can be achieved if the following characteristics are available:

a) ecological balance is available in the region, normal reproduction of the population is ensured, reserves to broaden and intensify the economic activity are available;

b) economic structure of the region allows one to provide effective employment of able-bodied population;
c) financial base of the region is sufficient for its self-development and for solution of social problems;
d) utilization of the natural resources' potential ensures long-term functioning of the region's economic complex;
e) external economic, social and political links of the region promote its economic growth.

Such a combination of economic, ecological, social, political and resource aspects of the development introduces principally new feature in the evaluation of results and prospects of the development of the regions. This makes one treat the possible strategies and alternatives in a new way.

Speaking about the new instances determining the position of Siberia on the threshold of the coming century, we should also note the integration of Siberian regions into the world economic system that can serve as a lever promoting a way out from the grave economic crisis.

It is obvious, that the Russia of today is not the best place to attract foreign investors and to develop external economic links in a large scale. We will not dwell at length on the estimate of general tendencies and prospects, but it is worth emphasizing that, of all large Russian regions, Siberia has always looked the most attractive for foreign investors due to the combination of the following advantages:

— unique in amount and variety fuel and energy resources, mineral and raw materials, timber and water resources;
— geographical and new geopolitical position of Siberia as the steadiest and largest region of the Russian Federation;
— powerful industrial potential which could be a solid ground for further reorganization;
— availability of skilled labor force in the base industrial branches and in military-industrial complex;
— unique potential of Siberian science is practically untapped.

It should be noted that it is precisely the combination of all these advantages that is of principal importance. However, it is required to begin the economic mechanism which will allow one to benefit from these advantages being too far uncovered.

When choosing possible ways to solve practical and strategical tasks of Siberian regions' development — "the model" for the economy's development — it is essential to use foreign experience.

However, we are not speaking about blindly copying American or German federalism, Swedish self-government, Canadian experience in development of the North. The specific character of Russia, and Siberia in particular, is too strong and the force of inertia and the heritage of the past are too strong for a foreign social and economic model of development to be automatically adopted.

But, a thorough investigation of foreign countries' experience in the realization of regional policies, development of local government, formation of financial resources of territories, development of the regions having extreme natural-climatic conditions, support of aboriginal population may prove to be of the utmost importance for the elaboration of our own "Siberian model" of the development. This model has to use thoroughly foreign countries' experience of many years (and of many centuries), where one should learn not only on achievements but also on errors.

The international collaboration of scholars from different countries is the most important and necessary stage of these studies. The economy of Siberia is currently a unique proving ground for the organization of the collaboration, where a twofold task can be solved:

— the acquisition of new knowledge, the development of the theory and methods of regional studies;
— the preparation of the ground to put the acquired knowledge into the practice of management and regulation of the development of Siberian regions, branches of industry and enterprises, into the system of local government, into the system of interrelationships between the regional and federal levels, into the system of the organization of long-term trade-economic links of Siberia with other countries and its integration into the world economy.

Therefore the task of considerably intensifying international contacts and joint economic studies of development of Siberia with foreign scholars are of moment and urgency. In solving these tasks a special place is occupied by the scientific projects being carried out by the Siberian International Centre for Regional Studies.

2. The Siberian International Centre for Regional Studies' Activity

The goal of the Siberian International Centre for Regional Studies (SICRS) is to make more active the regional studies in the Siberian Branch of the Russian Academy of Sciences (SB RAS) and to provide a scientific base for the integration of the Siberian economy into the world economy. SICRS has been established in SB RAS on the basis of the Institute of Economy and Industrial Engineering (IEIE). The founders of the Centre are SB RAS, the Ministry of Science, High School and Technical Policy of the Russian Federation. SICRS is an independent nonprofit organization and has an open institute status.

The Centre's activities will concentrate on the following basic directions:

I. **International research projects** carried out by the staff of the IEIE, SICRS, and other Russian organizations with the participation of interested foreign scholars and specialists. These projects are dedicated to theoretical issues of regional economics and interregional interactions; federal, international and interregional problems of Russia; Russian regional policies and the new regional economic mechanism in Siberia; the strategies for the development of Siberian basic industries; social policies in Russia and Siberia; ecological problems and economic evaluation of the mineral resources of Siberia; integration of Siberia into the world economy; conversion of Siberian defence-related enterprises; privatization and small business in Russia and Siberia; regional modelling and information systems, etc.

   The projects are commissioned and funded either by foreign governmental and non-governmental organizations, universities, foundations, and firms, or by Russian federal or regional authorities and other institutions.

II. **Training and retraining** of top and medium-level personnel for regional authorities of Siberia and the Far East. The Centre will invite leading foreign scholars and practical specialists, introduce standardized courses and business games widely used in business administration and management schools of other countries.

III. **Organization of business contacts** between foreign companies and Siberian enterprises, evaluation and independent expertise of various projects in the field of foreign economic activities in regions of Siberia, preparation of analytical reviews on the Siberian economy for republican authorities, as well as for foreign customers. This work can be performed with the participation of foreign specialists invited by the Centre.

IV. **Publication of an international journal** on theoretical and applied problems of regional policies, territorial development and management.
Beginning in 1994 the scientific journal "The Region: Economics and Sociology" issued by the IEIE will be published jointly with the Centre as an international journal both in Russian and in English.

V. **Organization of international study-tours**, symposia and seminars.

VI. **Participation in training** of postgraduate students, organization of scientific exchange of researchers. The SICRS can undertake on a contractual basis the organization of specialization training of foreign post-graduate students and their visits to Siberian regions and enterprises.

In fact, the Centre is an IEIE SB RAS segment for international relations and specializes in the organization and realization of international links, projects and programs of the institute.

The constituent conference of SICRS was held in March, 1992. Since the Centre has been actively working, it has acquired the status of a legal entity and has begun the realization of a number of international projects concerning the problems of Siberia. At present the following projects of IEIE financed by foreign funds are being realized through the Centre:

1. Structural Changes in Regional Economy under Transition to Market Relations (the Case of Kuzbass and Pittsburgh Region of the USA). The partner in the USA is Pittsburgh University, the grant for joint studies was awarded by United States Information Agency (USIA).

Comparative analysis of the two regions in Russia and the USA lies at the basis of the project. The Pittsburgh region was world metallurgical centre and lost its leadership in the post-war period in severe competition with the metallurgy of Japan and Germany but it managed to change its specialization to high technology production, in doing so the tasks of the environmental protection and personnel retraining were solved. The Pittsburgh experience is of the utmost interest for Russia from the viewpoint of the organization of the cooperation between science, local administrative bodies, and private firms in solving regional strategic tasks.

One of the painful points of present Russia is Kuzbass. It is a region of coal-metallurgic industry with acute economic, ecological and political problems. It is necessary to carry out economic and production restructuring of Kuzbass together with the solution of the social problems (social protection of the population, unemployment liquidation, personnel retraining) and the large expenses on environmental protection and the introduction of advanced technologies.

The foreign experience in the solution of the mentioned tasks is of utmost importance and the project's goal is to study and recommend it to the governmental and local bodies and administrative structures in order to get over the crisis in Kuzbass economy, to improve the structure of the economy, to use the mechanism of combining the interests of Russia, Siberia, Kemerovo region, different enterprises situated there, the group of entrepreneurs etc., thus realizing the program of social and economic reconstruction of the given region.

2. The project: "Institutional Strategies of Sustainable Development: the Development of West Siberian Resources". The US partner is Clark University, Mass. The grant has been awarded by McArthur Foundation.

The main concept of the project concerns the study of the influence of the processes in economic and political system of Russia (ownership change, the change in the methods of the states regulation of economic and ecological development) on the tendencies of the development in resource based and environmentally unsafe regions in Siberia. In other words, we mean a comprehensive analysis of economic, ecological and institutional issues of sustainable development after the example of the Kuzbass and Tyumen regions.

The mutual influence of the following processes is assumed to be studied: state regional policy — tendencies of regional development and self-government in the regions of Siberia — the processes of privatization, establishing share-holding enterprises, demonopolization in the basic branches of industry — economic and legal mechanism of resource utilization — mechanisms of regulation of ecological development and environmental protection in the regions.

In the process of work on this project, the study of the possibilities and the recommendations to governmental and local authoritative bodies on the regulation of regional development, resource utilization and ecological safety in the chosen "model" regions are also assumed. The general goal is to study concrete combinations of different forms and rights of ownership (including the rights on resources), legal regulations and administrative role of the state which promote ecologically friendly development in Western Siberia.

Concrete objectives of the project are:

— the study of the level and the type of degradation of the environment which is connected with coal, oil and gas production in West Siberia oil and gas complex and in Kuzbass;
— the estimation of the efficiency and applicability of various combinations of rights and duties for the regulation of the production and utilization of the resources in Western Siberia;
— the estimation of political and social acceptability and feasibility of various models of ownership and regulation in Russia and Siberia (here the American partners carry out comparative analysis of these processes in other countries);
— the preparation of the proposals on the realization and organizational-legal possibilities of the strategies for control over the resources on regional and local levels;
— the development of a general theoretical base for the analysis of the production of non-renewable resources on the basis of the concept of property rights and the strategy of environmental protection.

By now the first stage of the project has been completed and the preliminary materials, which will be discussed during the working meeting of main executors in the USA, have been prepared.

3. The project "Forecasting of Regional Development (the Case of Siberian Regions). The partner in the Netherlands is the Economic Geography Institute, Erasmus University, Rotterdam. The grant has been allocated by the Dutch Science Council.

The project is of comprehensive character and covers various aspects of regional development in Siberia and interregional relationships. A special emphasis is made on the problems of improving the mechanism of regional development and the methods of its regulation and forecasting, adjustment of the regions of Russia and Siberia to conditions of federated relationships under change.

The main objectives of the project lie in the development of methods of forecasting and analysis of the processes of Siberian regions' development in condition of transition to market relations. The project comprises the following basic issues:

a) "Forecasting the Development of New Resource-Based Regions of Siberia (the Case Study of the Lower Angara region)". Jointly with Dutch colleagues we will accentuate the study of methods for elaborating the program of development of region's resources, taking into consideration new economic relations and methods of management (suited to the conditions of the Lower Angara region). Particularly, Dutch scholars and practical workers will act as experts in the "Program for Development of Lower Angara Region" worked out by IEIE SB RAS. The monograph based on the results of the joint in-depth study will be published in the Netherlands in English.
b) "Regional Economic Policies and Development in Conditions of Federalism". The proposals for the improvement of the state regional policy regarding Siberia and of interregional economic and integrational relations of the regions of Siberia will be prepared. Realization of the above objectives presupposes the application of the achievements of modern economic science (equilibrium and quasi-equilibrium theory, theory of cooperative games, theory of federalism), economic-mathematical calculations;
c) "System of Informational and Analytical Support for Siberian Regions". The creation of economic tools is projected to evaluate the development of Siberian regions under different models and mechanisms of regional management. Analysis of behaviour of participants of market relations of regional economy will be carried out. "Full-scale" testing ground (one of the regions of Siberia) will be created for testing various models of regional management and the ways of transition to a market system.

In the fall of 1993 in the Netherlands the third Dutch-Russian seminar on regional economics took place as a part of this project.

4. The project "American and Russian Federal Systems: Basic Concepts, Experience and Future".

Among the great number of really topical and urgent problems of present-day Russia, there is one that seems to be the most important, acute and difficult to solve. This is the problem of creating a truly federal system in the state of Russia, one which would be the best in solving the questions of combining national and regional interests, state power and local government, common to all of humanities values and national peculiarities.

When creating the system, it is necessary to consider the experience of the former USSR, which in reality was not a federation with an effectively working system of shared power, sovereignty, and actively-working cooperation between the Centre and provinces and with actual elements of civil society. Territorial and national peculiarities of the USSR which influenced greatly all sides of economic and political life of this country were ignored and to a large extent contributed to the collapse of the largest unitary state in the world.

We should learn from the mistakes of the past using the positive experience gained in the history of common life of a large number of different nationalities. It is necessary to study thoroughly the foreign experience of the countries with federated state systems. This objective was pursued by the project and by the program teaching regional leaders of Russia the principles and experience of the American federal system. The given project was worked out by USIA, the Centre for the Study of Federalism, Temple University (Pennsylvania, USA), jointly with the Siberian International Centre for Regional Studies and IEIE SB RAS.

Originally, the project had been planned to be of a training character. IEIE SB RAS and SICRS were supposed to grant an opportunity to leaders of legislative and executive authorities of Russian regions to study peculiarities of how federal, state and municipal governments function, of the ways in which they co-operate, in methods of combining political and economic tasks of development, etc.

However, in the process of the project's realization it was decided that its main results should be the recommendations on expediating the application of whatever helpful or those elements of American federalism in Russia to federal and regional authorities (concrete questions, connected with formation of financial base of regions, ways and methods of their interaction, division of responsibilities on federal and regional levels etc., were very important). Besides, it was agreed that the main result of the program would be Russian-American monograph published both in Russian and English where the sections prepared by IEIE SB RAS and Temple University scholars and articles of the training program's participants would be included.

The present project consists of the following stages:

1. January, 1993. The seminar in Novosibirsk Akademgorodok for 25—30 representatives of local administrative bodies and Soviets of People's Deputies. Selection of 10 candidates for a tour of the USA.
2. May, 1993. Intensive seminar in Philadelphia for the ten regional leaders. Acquaintance with the experience of cooperation between federal and local authorities in the USA, organization of local government. Study tours to the capitals of the states. Selection of the candidates for internship.
3. June, 1993. The internship of five Russian regional leaders in local governments and communities of the USA.
4. July, 1993. The trip of the second group to the USA for a short (two weeks) program of training.
5. September, 1993. The final seminar in Novosibirsk for the whole group (25—30 participants).
6. Preparation of a version of the monograph (collected works) "American and Russian federal system: basic concepts, experience and future" in Russian (to be published at IEIE) and in English (to be published at Centre for the Study of Federalism, Temple University, USA).

In the framework of the given project at the present time the tour to the USA of 5 governors of the most important regions of Siberia is being prepared to study forms and methods of regional policy implementation, local government, interaction of federal and local authorities, activity of various interregional organizations (National Association of Governors, the National Conference of State Legislatures and The National League of Cities etc.).

Thus, today we can speak about the first success of international research (and training) projects being realized in the interests of Siberia jointly with foreign partners. Financial support (grants) for these projects received from governmental bodies of different countries and from charitable foundations is of great significance. It promotes not only the possibility of intensification of international scientific relations and "internationalization" of economic studies in Siberia but to a large extent influences the preservation of the potential of Siberian economic science when new incentives for scientific activity (including material ones) are created.

Presently good perspective for the joint organization of foreign scholars research projects are available. The study of development of frontier regions of Siberia is very promising.

The project may be realized jointly with Paul Valery University (Montpellier, France) with the participation of American and Polish scholars. The significance of the frontier regions of Siberia grows as far as it is connected with new the geopolitical situation of Russia and Siberia occurring from the break-up of the former Soviet Union. Frontier areas in foreign countries presently realize a principally different function than they did earlier, being outposts but in fact playing the role of typical outlying regions with low-developed economy and nonprogressive economic structure.

Currently, frontier regions have begun performing new functions (including the ones being redistributed from the centre into these regions) which is connected with the openness of the borders, intensification of international integration. Today the openness of borders and the economy in different countries removes the stigma of provinces attached to frontier areas, the latters are becoming powerful epicentres of newly arising international economic, cultural and transport relations.

One of the authors of the present paper, Prof. S.Savey, undertook an international investigation of tendencies and prospects for the development of frontier areas studying the ones in Poland and Ukraine and Tunisia. Originally advanced hypothesis about changes of functions of these regions and directions of their development have been fully confirmed. And now it is of great significance to carry out a joint comparative study applying it to the regions of Siberia. The Lower Angara Region in East Siberia has been selected as a testing ground for the fulfillment of the joint project.

By and large, the recent conference, held in SB RAS together with NATO Scientific Affairs Division and other international organizations, proposed, as of priority, the topics for social-economic studies which are the following:

1. Interstate economic relations within the the Commonwealth of Independent States: analysis and forecast.
2. Regional policy in Russia and new role of Siberia: main directions of structural, investment and social policy of Siberian regions.
3. Global changes and Siberian frontier areas: new perspectives.
4. Integration of Siberian regions into world economy.
5. Industry of Siberia: possible strategies and evaluation of their socio-economic consequences.
6. Conversion of the defense industry establishments in Siberia (regional aspects and socio-economic consequences);
7. Development of free enterprise and small business in Siberia.
8. Social monitoring of economic reform in Siberia.
9. Problems of demography and migration of population.
10. Comparison of the principles of federal structure in countries of the West and the East.
11. Institutional strategies of sustainable development of Siberian regions, legal aspects of production of natural resources.

The development of above topics and projects jointly with foreign scholars, financial support of the works by governmental and nongovernmental bodies of different countries, foundations and universities will promote both development of social and economic studies in Siberia and the support of Siberian economic science and allow foreign specialists to acquire professional expertise, forecasts and estimates of the development of Siberia, its regions and enterprises and to use the results for significant intensification of integration into world economy.

♦

THE BAIKAL REGION SOCIAL ECOLOGY ISSUES

A.K. TULOKHONOV, B.O. GOMBOEV and T.B. BARDAKHANOVA
International Scientific Centre of Socio-Ecological Problems of the Lake Baikal Basin, Baikal Institute of Nature Management of SB RAS, Ulan-Ude, Russia

ABSTRACT. The main goal of the Centre of Socio-Ecological Problems of the Lake Baikal Basin is to work out the mechanism of the Baikal watershed sustainable development. This complex scientific problem consists of the following directions: paleoecology, assessment of the health of population, ecologically safe technologies, agriculture, problems of traditional wildlife use, landscape planning, ecological education.

Keywords: social ecology / paleoecology / health of population / sustainable development / agriculture / traditional wildlife use / landscape planning / ecological education.

The ongoing political and economic changes in Russia have touched all facets of social life of the country. The possibility has emerged for a broad scientific exchange and international cooperation which includes the establishment of International scientific centers within the context of the Siberian Branch of the Russian Academy of Sciences.

Two international centers of learning hosted by the Limnological Institute in Irkutsk and by the Baikal Institute of Nature Management in Ulan-Ude, are addressing the problems in the natural sciences and social ecology respectively, for the Baikal region.

The main goal of research of the Centre of Socio-Ecological Problems of the Lake Baikal Basin is to work out a plan for sustainable social and environmental development in the Baikal region. Practical implementation of this complex scientific plan can be carried out by a series of interdependent fields.

1. Paleoecology. This field deals with the study of ancient man's habitat on the banks of Baikal, and his method of nature management. By means of a wide range of geological, archaeological, botanical, and radiocarbon dating methods one is able to measure the climatic changes of the earth's northern hemisphere from the tightly fixed sedimentary deposits in the Baikal. The study of ancient man's art, characterized by rock paintings and ritualistic and household items is of special interest. The study of paleoecological aspects creates a basis for determining the fundamental state of the environment and the distribution of natural components over prolonged observational periods.

2. One of the most important ways of determining the ecological well-being of any territory is by assessing the health of its people, especially in industrialized areas. In order to solve this problem an analysis on the medical statistical material and basic indicators was conducted in all

V. A. Koptyug and J. Klerkx (eds.), Science Policy: New Mechanisms for Scientific Collaboration between East and West, 209–214.

large populated centers. It showed average and abnormal indicators by means of positive and negative signs. The difficulty in coming to a final conclusion is due to the impossibility of identifying a clear dependence of socio-economic factors on environmental conditions.

At first glance, we can only theorize that the rate of disease in industrialized centers is higher. However, final conclusions require more serious investigation.

3. The health of the population is directly related to the standard of living, which in turn is effected by economic factors.

The difficulty of the problem with a sustainable rate of development is caused not only by the increased level of efficiency in production, but above all, by the need for ecological safety and structural reorganization in favor of maximum adaptability to an area's ecological conditions.

In the field of industrial development as it presently stands, priority is placed on development of those branches which do not require a large investment, will promote the most effective use of the geographical location of the centrally located Baikal region, and at the same time have a fairly developed transportation network with outlets in Mongolia, China and other countries of the Asia-Pacific region, and a great potential for recreation and an abundance of natural resources.

Such an approach does not dismiss the existence of other types of industry, but due to the unique ecological limitations of the Baikal region a greater expense is required for measures taken by these enterprises to preserve the environment. The development and implementation of ecologically safe techniques, in particular the use of plasma coal burning for heating can possibly change the outlook for the development of those branches of production which traditionally pollute the environment in the region under review. The introduction of ecologically safe and environmentally friendly technologies can effect the scope of development of the lumber industry, which in the past has been a significant part of the economy in this region.

4. Farming is one of the leading branches of the region's economy. Livestock management has traditionally been the main vocation in the Baikal region. The prevailing form of livestock management has been nomadic. In farming priority was given to the production of livestock feed and cereals which were the most adaptable to severe local conditions. Such a system of land-use was the most economically beneficial, and preserved the viability of the land.

The subsequent unification of agricultural production has had negative effects on the condition of the agricultural sector of the economy. The plowing of virgin and long-fallow lands in the 50s reduced the area of fields and pastures which in turn undermined the food supply necessary for livestock management.

That decrease of fields and pastures coincided with the sharp rise in the number of sheep which are more detrimental to the land compared to other domesticated animals. Over the period of 100 years the area of pasture land has decreased by 3 times. At the same time the burden on the pasture land has increased by 6 times, due to the present high rate of erosion of the land and the loss of the initial bioresource potential of the land.

The consolidation of rurally populated areas and the concentration of agricultural production around them, have brought about a sharp rise in the anthropogenous burdens on the environment. Traditional forms of livestock management have gradually died out: productive distant pastures are used less often, which leads to a decrease in nomadic forms of livestock management. At the same time new techniques in agricultural production have not been fully developed because there is not an adequate supply of natural vegetation, and due to the lack of necessary sewage facilities and specially equipped storage houses for storing pesticides and mineral fertilizers, the anthropogenous burden on the environment has increased.

5. Presently structural reform is needed in agriculture: above all, this means the renewal of traditional forms of farming and nomadic livestock management. With this in mind, the potential for livestock management in Buryatia has been explored. Work has begun on breeding indigenous animals that are already suited to year-round subsistence on these pastures. An experimental farm was set up in which 570 head of shaggy wooled sheep where purchased in China and were acclimatized here in Russia. Yaks have been brought to the steppe areas of Buryatia, and in the near future the acclimatization of camels is planned.

The methodology for the selection and preservation of animals who live a nomadic lifestyle has also been worked out.

6. Landscape planning of specifically protected natural territories.

It is common knowledge that the establishment of a network of protected territories involves a search for an optimal balance between natural ecosystems and the anthropogenous areas. In order to realize these goals, we are making provisions for the terrain of the Tunka national park by identifying functional zones, which show a possibility for various uses of their area. The park's functional zones are thought of as objects which are located and developed within an administrative district, and because of their structural peculiarities, technological processes and functions they perform make specific demands on nature's resources and are characterized by concrete amounts of consumption, reproduction and pollution of the natural components.

The territory of the Tunka national park is divided into 8 functional zones:

1) the wildlife reserve zone which is broken down into two subzones — the mountainous taiga and the bald (glacier) peaks;
2) the zone of limited use;
3) the zone for the preservation of the fauna;
4) the zone of traditional economic activities, which is broken down into three subzones — lowland, mountainous taiga forests and mountainous taiga pastures;
5) the service zone;
6) the agro-park zone;
7) the zone for recreation and the lumber industry. These zones are shown on the map scheme 1:2000000.

For each functional zone a method of utilization is developed in order to meet the needs of agricultural, environmental, and recreational concerns.

The main objective of the "wildlife reserve zone" is the preservation in a natural and undisturbed state the diversity of all natural components, including the characteristic and unique nature of plants and animals. Here agricultural activities are absolutely prohibited. In the zone of limited use there are designated areas where hiking paths have been laid down and where mountain recreation is permitted at designated altitudes. The "Badary" national game reserve has been designated as a zone for the preservation of the fauna. The zone of traditional agricultural use has been set aside for the preservation and reintroduction of traditional approaches to the economy and its relation to the use of renewable natural resources: wildfowl, fish, wild berries and nuts, as well as developed livestock management. The main goal of this agro-park zone is to supply food to the local population and visitors of the park. Here there is limited use of natural fertilizers and other substances. In the near future further studies which will help to divide the zone for recreation and the lumber industry will be conducted.

By and large the work on the creation of a system for landscape-economic partitioning of territories in the Baikal region will continue in the future. It is being suggested that these studies should be extended to the Barguzine valley, Buryatia and other territories in this region.

7. The solving of all of these pressing problems and the realization of these scientific recommendations can be made possible only through the use of this mode of thinking, which views nature as an object of use and a provider of certain products. Therefore it is necessary to give top priority to an ecological upbringing and education for the coming generations in accordance with national traditions and customs; on the basis of the belief in the inseparability of man, society and his generation. This generation will recognize that the Baikal is valuable not only to those people who live along its banks but belong to all of mankind. At this stage, the first experiment on ecological upbringing and education in Europe and in the East has been summarized. A library of text books from Germany, England, the USA and Canada has been established. We are developing our own books and games. It was very useful to us to study the ecological upbringing and education of the Indians of the US and Canada and the concrete approaches which they developed in order to tackle a given problem.

It is important to emphasize once again the special significance of developing international ties in the field of science and business because in this way we are following the course of more developed countries. And the experience which they have gained will undoubtedly help us to shorten the journey towards our ultimate goal which is the preservation of the Baikal and the environment as a whole.

Studies on four projects have proved to be more effective concerning the activities of our International Scientific Centre of Socio-Ecological Problems of the Lake Baikal Basin.

The first project. The authors of the project are "Davis Associates", the Russian Academy of Science, the US Civil Initiatives Center (USA), and the International Scientific Centre of Socio-Ecological Problems of the Lake Baikal Basin.

Work on the "Comprehensive Program for Land Use for the Russian Territory of the Baikal Basin", was completed in 1993. The Program and map of land distribution have been discussed in public meetings, meetings with scientists of different state institutions in Buryatia, Irkutsk and the Chita regions and with experts of the international community who familiarized themselves with the program. A series of projects such as, the creation of a communication system, organizing model farms, recreation zones and the revitalization of separate plots of land bordering Lake Baikal and other areas, has been proposed for implementation.

The second project. "Environmental and Cultural Conservation of Inner Asia Regions" is in the state of implementation. The above mentioned region is one of the centers of civilization. It possesses unique natural conditions, economy, culture and religion of the population. Inner Asia includes Buryatia and Tuva which are within Russia's borders, Mongolia and two autonomous regions of China: Inner Mongolia and Sinidzyani-Uiguryan autonomous region. Cambridge University aided by the U.S. McArthur Fund grants is the coordinator of this project (the scientific head is Dr. C. Humphrey). The chief executor of this project model is BIRP of SB RAS, and as a standard site Buryatia was selected. In relation to this, the natural scientific and social components are being worked out at BIRP.

The main goal of this project is to study the dynamics of nature in Inner Asia under the influence of various socio-cultural processes which have been taking place during the 20th century, and to forecast possible changes relating to the evolution of forms of nature and the political reorientation of former socialist countries.

Research methods include a retrospective ecological analysis of the history of development of agricultural economy in a region for a period of more than one hundred years and the collecting statistical data on the state of farm management by means of farm councils in Russia and their counterparts in Mongolia and China (of which there are more than 500). During the eighties and nineties the dynamics of nature have been analyzed through data received through satellite photos (of various scales and spectrums). The results from deciphering the satellite photographs are transformed into a series of maps on natural environmental changes and supplemented with materials from field investigations on the degree of viability of steppe terrain.

By and large, the information can be divided into 5 levels:

Level I includes information referring to general aspects of agricultural land use in Inner Asia. This reflects the structure and total head count of livestock, and the number and nationality or the inhabitants. The cartographic data was collected to 1:5000000 scale. From this body of data isolinear maps on the density of the human population, cattle, its structure and dynamics in accordance with the indexes of the third level (for the Russian area) and those of the fourth (for the rest of the regions) will be drawn up.

Level II includes statistics concerning large regions in Inner Asia. The main part of present economic development and conditions of the environment have been singled out. More detailed material was gathered on Buryatia and the Baikal region. The results of the research will be reflected in a separate series of maps to 1:2000000 scale.

Level III includes the investigation of concrete units in the region of the Barguzine valley, the south of Buryatia and Chita region and Central North of Mongolia. Natural and anthropogenous processes and phenomena which influence seasonal and territorial land use are of the greatest interest.

Level IV is the level of village Soviets (administrative units of the farm council type). In 1993 a large scale research expeditions were conducted with representatives from a British youth organization, "Raleigh International".

Their work entailed investigating the Barguzine valley pasture conditions. The inhabitants of the area were asked to fill out specially prepared forms. Seven village councils and sixty-five pasture lands have been inspected in all. The number of control measurements is 50000. From these data a measure of the general viability of the pasture is arrived at. This enables us to determine a possibility for various breeds of cattle to subsist on it and also the measure of optimum load that a pasture can bear.

Level V presents a socio-ethnographic examination of eighteen families according to a specially formulated 600 page questionnaire. The above mentioned families are differentiated by social status. They make up all basic social, professional, and age groups of the population. The inquiry took place in the Argada village council in the Kurumkan district of Buryatia. Analogical inquiries were conducted in all regions of Inner Asia (for a total of 9 sites). This will help us in the future to determine the attitude of concrete societies toward the transformation of nature and the state of the cultural environment under changing conditions.

Thus, the formation of a data base for the project was conducted in this manner in order that all levels will be closely bound and logically complement each other. The results yielded are the basis for several conclusions:

1. During the past one hundred years Inner Asian steppe agrolandscapes have undergone significant changes which above all, manifest themselves in the development of cultivated crops, uncharacteristic of this territory.

2. In all of these areas the rapid development of the process of erosion which is linked to the high concentration of livestock is being observed.
3. It is necessary to change the structure of agriculture in spite of obvious contemporary economic loss in favor of those forms which are adaptable to local environmental conditions.
4. The structural reorganization of the economy in these areas should be accompanied by reorientation of the goals of society and the revival of practices of traditional use of the environment.
5. Interesting results can be obtained through analysis of socio-cultural living conditions of various nationalities in various regions in Inner Asia. However this is a subject for further study.

Undoubtedly all of these conclusions are preliminary and will eventually be revised. However, the general foundation of this work has already been created.

The third project concerns landscape planning of specifically protected areas of the Baikal region and is in a preparatory stage. The German Ministry of the Environment, which protects nature and monitors the safety of nuclear reactors, is the financial and scientific partner abroad. Also the International Secretariat for the Preservation of Nature (EEC) is interested in the development of this project.

The final objective of the project is organizing a system of specially preserved natural area in the Baikal region on the basis of landscape planning which in its turn is a reasonable and effective means of keeping up the natural balance, natural resources and protecting the already existing local landscapes.

The main documents for the preparation of the project are to be signed in December 1993.

The fourth project is particularly specific for our region and aims at organizing a system of adaptive use of nature in the Baikal region, taking into account its historical, national, cultural and religious traditions. At the first stage we chose two categories:

The first is resources. It is directed towards researching the principles of reconstructing a traditional agricultural structure which is adequate for the natural territorial conditions. Different scientific organizations of Mongolia (The Mongolian Institute of Stock Breeding Economics), Inner Mongolia (The Inner Mongolian Economics Institute, The Teachers Training University and the Inner Mongolian Geographic Society) as well as the Chinese Scientific Research Society of Land Economics are being encouraged to cooperate.

The second category is directed towards the creation of a continuous ecological educational system and study, with due regard to native people's traditions and also the experience of European and Eastern countries, and the Indians of the USA and Canada.

Agreements for cooperation in this research with specialists in Germany, Canada and the USA have been concluded.

In concluding, it should be noted that the first steps toward implementing the directive for the research of the International Scientific Centre of Socio-Ecological Problems of the Lake Baikal Basin were realized at this point, in the development of the abovementioned scientific projects. This enables us to hope that a real mechanism for international scientific cooperation, which enables us to exchange information on the environment and pass on state of the art technologies and experience will be created. These techniques and experience can be applied in different regions of the planet.

♦

PERSPECTIVES AND PRELIMINARY RESULTS OF ARCHAEOLOGICAL INVESTIGATIONS OF THE SOUTH-WESTERN ALTAI (THE UKOK PLATEAU)

V.I. MOLODIN
Altai International Centre for Humanitarian and Biospheric Research,
Institute of Archaeology and Ethnography of SB RAS, Novosibirsk, Russia

ABSTRACT. The four year investigations at the Ukok Plateau have resulted in the elaboration of the general concept of historical and cultural development of the region during Holocene. It is argued that the region was populated starting as early as the Stone Age through the Bronse and Early Iron Ages up to the Medieval Ages. Excavations revealed the vivid sources on material and spiritual culture of the highland inhabitants. Most interesting are the burial mounds of the Pazyryk culture. Due to the peculiar construction character of burial chambers lenses of ice originated inside the graves, thus preserving unique wooden, textile, and felt objects, as well as mummified human bodies.

Keywords: pazyryk culture / petroglyphs / turkic time / bronze age / frozen tombs.

The Ukok Plateau stretches upon the vast territories in the South-Western part of Gorny Altai bordering Mongolia and China in the south and Kazakhstan in the west. This highland plateau is 2500 m above sea level, it is surrounded by snow-covered peaks and high mountain ridges and is hardly accessible even nowdays. There are only few narrow valleys that hardly provide passage into the area. The Ak-Alakha and Bukhtarma mountain streams originate in the plateau. Given the severe climatic and environmental conditions of a highland boreal steppe, one can understand why the area and archaeological sites on the plateau were hardly affected by any anthropogenic factors; due to that fact the area has practically retained its protoplastic outlook.

At the same time, however, favourable potentials for cattle rearing in the winter have been attracting cattle-breeders for centuries beginning from the Early Bronze Age till present times. Due to the reasons enumerated below, the area in question has promising research perspectives.

1. Due to its geographical position the Plateau was inhabited by the tribes with highly expressive and original culture, starting probably from the Stone Age up until now. This fact allows us to reconstruct the whole picture of the ethnocultural development in this region for at least 7 millenia. As the area had occupied the bordering geographical position for some time it made an impact on the development of both ethnic and cultural phenomena.

2. Extreme climatic and environmental conditions of the area favoured the emergence of lenses of ice in some burials of the Early Iron Age leading in its turn to the perfect preservation

V A Koptyug and J Klerkx (eds), Science Policy New Mechanisms for Scientific Collaboration between East and West, 215–222

of offertory goods made of textile, animal skin, felt, wood, an maintaining the conditions for mummified bodies preservation.

3. Weak anthropogenic effects on archaeological sites resulted in their present undamaged state. The fact increases the representative value of the revealed material. Archaeological investigations at the Ukok have already been conducted by two research teams in staff of the Institute of Archaeology and Ethnography of the Siberian Branch of the Russian Academy of Sciences for already 4 seasons now. A Comprehensive Project of research of various archaeological sites, including grave yards, places of cults, settlements, petroglyphs, has been elaborated on.

Experts in different disciplines, namely in geomorphology, geography, paleontology, anthropology, dendrochronology, etc., are participating in the project.

Investigations are conducted in close collaboration with many foreign scholars, with researchers from the Royal Museums of Art History in Belgium, from Harvard University, U.S.A., the City Archaeological Centre in Zurich, Switzerland. Excavations at the Ukok were visited by specialists from Great Britain, Germany, South Korea, Japan, etc.

General results of the field works at the Ukok are led in two major directions.

Fundamental research. Research works resulted in an elaboration on a scheme of cultural and historic development in an area which would encompass the whole period of Holocene. In general the scheme goes as follows (Fig. 1).

The earliest site dating back most probably to the Stone Age (in particular, to the Mesolithic and to Final Upper Palaeolithic), discovered at the plateau is the rock art site of the Kalghuta Mine location. The sketches of petroglyphs represent images of horses, bulls or bisons (Fig. 1), done in a specific manner that allows them to be placed among the most ancient Mesolithic or Upper Palaeolithic rock paintings in Eurasia (1). This site of rock painting is currently the most ancient one discovered in the Altai.

The lack of raw material of high quality for lithic tool production together with the specific geomorphological situation makes the search for the Early Stone Age sites highly dubious.

The Early Bronze Age is represented by the burial complexes of the Afanasievo culture (end of IV — beginning of the II millenium B.C.) — (Fig. 1 3—11). It was a European population inhabiting mountain regions of South Siberia including the regions of the Altai, Khakassia, Tuva, of Western Mongolia and North-Western China. The interred bodies were laid on their backs, the legs were flexed with knees up, with heads oriented to the west or south-west. The burials were summoned by small construction made of pebbles intermixed with earth, in a form of concentric circles — ramparts. The typical inventory of grave goods included vessels, lithic tools and decorations made of bone (Fig. 1 3—11).

An isolated construction Bertek-56 can serve as an example of an Advanced Bronze Age site (Fig. 1 12—19). This site resembles a stone rampart with a diameter of 14 m with a pile of stones in the centre, with obviously peculiar constructive features. At the process of excavations, remains of five fire-places were found. In the centre of the burial construction two tombs were found. An inhumation was made at the level of a bed ground in a stone box resembling a cist with two long walls. The tombs were covered with 2 or 3 massive stone caps. Both burials are of a secondary type. The skeletal bones are laid in an anatomical order though some of them are missing. An inhumated body is laid on its side in a stretched pose with the head to the east. In the process of cleaning of fireplaces, potsherds were found belonging to three different vessels which allow to precisely date the whole construction to the period of the

Advanced Bronze Age. Analogues to the pottery can be found in a cult place in the middle reaches of the Kucherla river (2). The sites seem to belong to the new unknown culture of the Altaic Advanced Bronze Age located in its southern or south-western part and genetically affiliated to the Afanasievo culture (3).

A stone monument in the form of human head is most likely to belong to the same time period. The head was used secondarily for a construction of a barrow of the Early Iron Age burial mound. At any rate stylistically similar objects can be traced in contemporary culture of Okunievo at the Minusinsk hollow (4).

The rock drawings of a bull and a local deer discovered at the Kalghuta Mine site can be dated back to the Bronze Age. These drawings represent the side views of animals of rather large size, the style is distinguished in the ancient art of Gorny Altai as a separate cultural and historical type (5). The sketches of the kind have affiliations in the art of Okunievo culture on the territory of the Western Sayany mountains (6), in the contemporary cultures of the Tuva territory (7). The main body of Altaic rock paintings of this type are likely to belong to the Karakol period of the Advanced Bronze Age, the origin of this tradition, however, unddoubtedly stemmed out of the Afanasievo period.

Sketches in geometric style from the sites Morena I and Mazy-Bulak (Fig. 1 17—19) can be dated back to the Advanced or the Late Bronze Age. The sketches in similar iconographic manner are reported from the Bronze Age sites in Central Asian Republics (8) suggesting the invasion of the given territories of Gorny Altai by some groups of people from the West. The Late Bronze burial and settlement complexes are still to be found at the Ukok.

Noteworthy also are the Early Iron Age sites of the May-Emir culture of VIII—VII centuries B.C. (Fig. 1 20—29). Several burial mounds of the period were investigated, the most interesting among them is the Ak-Alakha-2. The barrow was constructed of huge cobbles, having the form of a flat pavement some 40 m in diameter. After its deconstruction a sophisticated structure was observed in the form of a rounded stone rampart with some annexes geometric in form and two rows of "deer stones" by its side. One of the "deer stones" was placed in the centre of the barrow. Though intruded, the burial chamber revealed the remains of the common inhumation of several human bodies and horse corpses. Included in the burial items were the stirrup-like bits of a horse bridle, and decorations providing solid grounds for dating the whole complex, and golden strips, etc. Associations of the deer stones to a burial complex with datable burial items is an invaluable phenomenon as it provides the basis for understanding these miraculous, monumental pieces of art.

Synchronous to the above-mentioned site is the Bertek rock art location, with depictions of camels, deer and wild goats composing some pictures in a manner typical of the Scythian period.

Starting from the VI century B.C. the May-Emir culture is substituted by another even more expressive one, the so-called "Pazyryk" culture (Fig. 1 30—48), which is spread over the whole territory of the highlands, and widely-known due to the works by S.I. Rudenko, M.P. Griaznov, V.D. Kubarev, A.S. Surazakov etc.

At the Ukok, excavations of the burial mounds of common nomads were entirely similar to the burials reported from elsewhere in the Northern Altai. The dead were interred in rather deep tombs, a wooden box was placed at the bottom. Single interments are the most typical, but there are some common burials. The dead was accompanied by a horse. Offertory goods included vessels, bronze and iron pieces of weapons; horse trappings, decorations were also found.

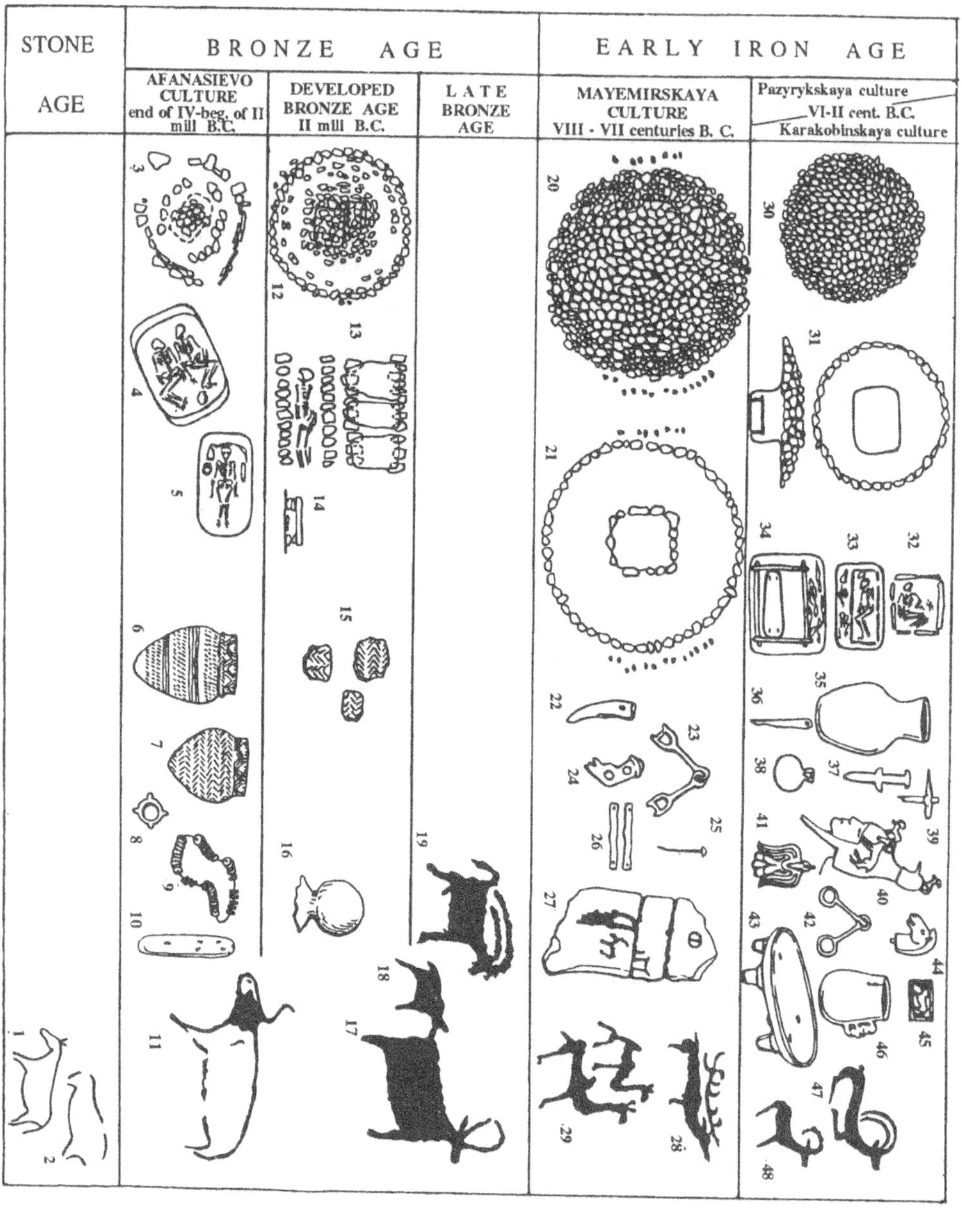

Fig. 1.
Correlation of the ancient cultures of the Ukok Plateau (1—48).

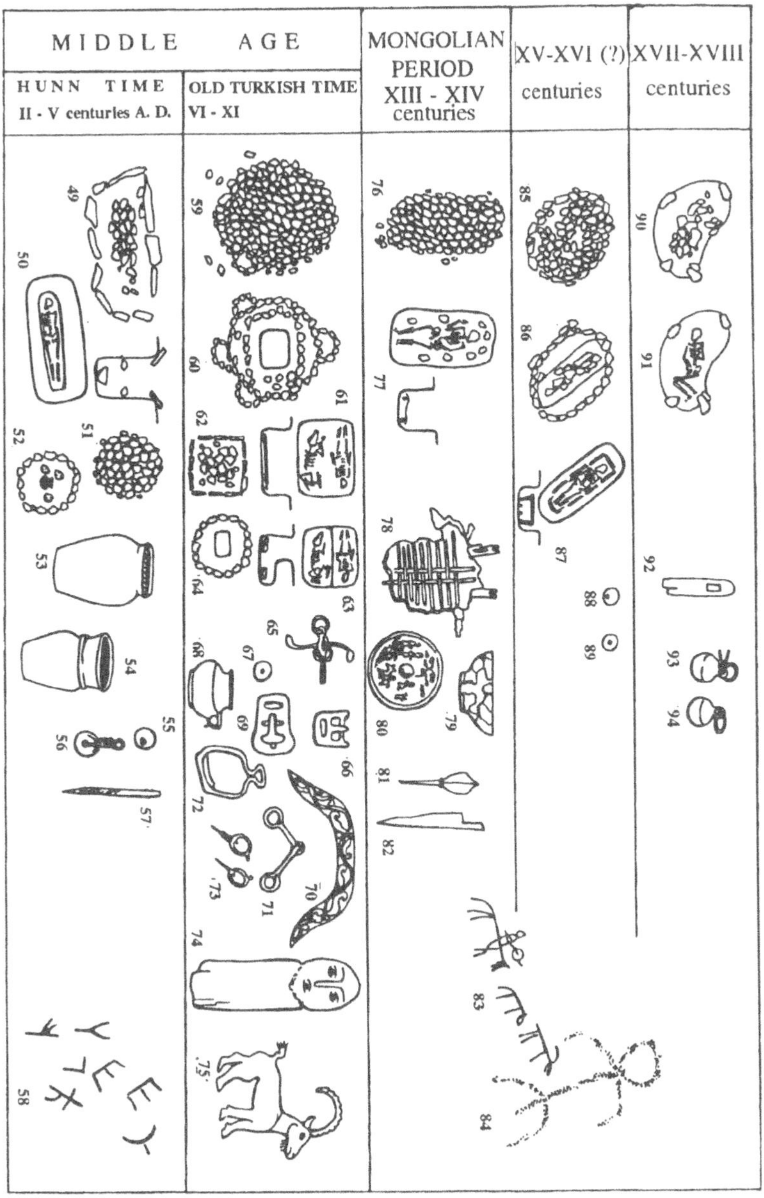

Fig. 1.
Correlation of the ancient cultures of the Ukok Plateau (49—94).

Investigations of two burial mounds of the Ak-Alakha 1 and 3 sites, probably of the middle class nobility, revealed very expressive material. The tombs turned out to be unintruded.

Due to the covering of the burial chamber by ice lenses some highly artistic objects made of wood, leather, and felt, were preserved as well as human remains and horse corpses (10).

Materials of excavations not only reveal objects of mobil art and handicrafts, but also provide solid grounds for realistic interpretation of economic ways and daily routine of ancient Pazyrykers as well as of their spiritual world. As for the preserved human and animal remains their discovery cannot be overestimated as material for paleogenetic research.

Some principally new information became available due to the investigations in the visible surface parts of the burial constructions of various types (11).

At the Ukok several burials of the Karakol culture which is synchronous to the Pazyryk one, have been investigated (Fig. 1 32). This culture is characterized by some particular features of mortuary practice, for instance, simpler construction of the visible part of the burial mound, and placement of the dead into the stone box (12). The Karakobiners seemed to come to the Ukok after the Pazyrykers. This is suggested by the fact that several secondary occupants clearly of the Karakobin culture were found in the Pazyryk interments.

A major portion of rock paintings at the Ukok belongs to the Early Iron Age. As a rule these are small groups of sketches on shale cobbles, which were smoothed by a glacier, or on the cliffs of the right bank of the Ak-Alakha. All the drawings have one and the same typical character. They represent side views of various animals, the most popular image being that of an ibex. Similar sites are reported from elsewhere in the highlands of Central Asia.

The grave yard Kaldjin-6 can be attributed to the Hunnic times. Some ten interments were investigated, where the bodies were placed into deep graves. The surface part of the construction is represented by a rectangular pile of stones with a rather deep depressions in the centre, resembling that of a cist. The whole construction is oriented along the line east-west with slight deviations. The body is laid on its back in a stretched pose with a head to the west — north-west, rarely — to the east. The bottom of the grave was laid by the wooden boards, sometimes more sophisticated wooden constructions were made. Sometimes the dead was laid into a wooden coffin chiseled out in the whole log of larch. The dead was accompanied by not very rich but mpressive inventory. A discovery of a massive jade bead and a pendant made of cornelian in a golden wire suggests a Chinese origin and allows to date the whole burial complex to the Early Hunnic period.

Concerning the Ancient Turkic period (VI—XII centuries B.C.), there are several burials and sacred places that were investigated at the Ukok. Two variants of burial practice were distinguished. The first one is constituted by a rather complex surface construction of stones forming a hexagonal rampart with linkages and attached circles that became visible after the above-laying stones were taken out of the pile. The grave pit is divided into two parts by a wall, the first chamber is occupied by a human, the second one — by an accompanying horse (Fig. 1 59—75).

The sites of the second variant have simpler surface constructions consisting of a single round ring of stones limiting the tomb. As for the grave pit, it is more complex, with the second chamber going to the side from the main one and separated from it by the vertically placed slabs. A body was placed in a second chamber, the main one was occupied by a horse. The common feature for both variants of the burial practice is an orientation of a body with his head to the east, and an opposite orientation for the horse.

A rich variety of tools was discovered in the Turkic tombs including labour tools, decorations, household objects and a horse harness. Noteworthy are the most impressive silver vessel from the site Bertek-34, and the saddle with the on-laid ornamented bone parts (from the burial 1 of the Kaldjin 7).

In general the Turkic sites at the Ukok can be attributed to the full flourish of the Kurai culture (middle of IX—X centuries). The emergency of the tombs with side chambers suggests the probably greater influence of the Uighur Caganat upon the ancient Turkic population of the Altai (17).

Interments of horses accompanied by a specific inventory as was found at the Bertek-27 site should be dated back to the X—XI centuries. Several stone hedges typical for the Ancient Turkic period were found at the Ukok. They have a typical rectangular form sometimes assosiated with an anthropomorphic sculpture. One of the sculptures of the kind was found in the Kurtunghuntas valley by Dr. N.V. Polosimak.

Rock drawings of the Ancient Turk period are casually found at the sites of various periods. They are usually engraved on the rocks in a perfect, realistic and dynamic manner.

The discovery at the Ukok of the sites of the Mongolian times are especially significant as they are poorly studied not only at the Altai, but elsewhere in the Central Asia (Fig. 1 76—82).

An isolated stone construction was investigated. The construction had a form of an oval pile of cobbles covering the grave pit. The body was laid on its back in a stretched pose with his head to the north. The chest was covered by a specific peculia piece of armor woven of leather strips and twigs. Armor of a similar type is well known in China, where it was still used in ethnographic times (18). At the same tomb a beautiful Chinese bronze mirror was found, it was ornamented in the back by a mythological scene. The object is dated to XIII—XIV centuries A.D. (20). Noteworthy is the fact that in the process of excavation of the Pazyryk burial mound the author used to find (in the upper layers of a stone pile) pieces of armor and porcelain ware of Mongolian times. This phenomenon is obviously connected with some unknown cults spread among the Mongolian population of the Ukok, for which purpose burial mounds of the ancient inhabitants were used.

Several burials can roughly be dated to XV—XVI centuries A.D. The burials usually have the form of an oval construction oriented along the line NW-SE — (Fig. 1 85—89). After destruction of a stone pile an oval hedge can be observed limiting the grave pit. In a small pit a box of stones was constructed as a place for a body. A body was laid on its back in a stretched pose with its head to NW. Offertory goods are scarce, only several cornelian beads were found.

Of interest is the burial site Bertek-9. It was made in a small pit and was covered with a pile of stones (Fig. 1 90—94).The body was laid on its right side in a flexed pose with its head to the west. The characters of the burial ritual, regardless of some individual traits, connect it with mortuary practices of the Kazakh tribes inhabiting the Southern Altai, the Ukok included. Archaic pre-islamic traits present some controversy to the identification. But according to A.V. Konovalov, an expert in the problem, some pre-islamic traits were preserved in a mortuary practice of the Southern Altaic Kazakh population which is free from any affect of Islam; which is likely the result of isolation of this group from the general ethnic bulk (22). This supposition is proved by a dead body's orientation along the line W-E-facing the south (i.e. facing Mecca), as was typical for the position of the dead of the Kazakh of the South Altai, Western Kazakhstan and the Bunei Horde (22).

Characteristic bronze buttons found in the tomb were exported from Russia. As the penetration of Russian exported goods into Asia took place not earlier than in the XVI century,it was most intensive during the period of XVII—XVIII centuries.

In rock art, the Ancient Turkic tradition is followed by petroglyphs of Paleoethnographic time. They are engraved and differ from the ancient ones by slipshod and schematic manner and by an absence of patina. Rock art of the Middle Ages and ethnographic times is known elsewhere both in the Altai and in Mongolia (23).

All of the above-mentioned works have outlined the general idea of historical developments at the Ukok, as in the South-Western Altai for several millenia. The outline is obviously full of lacunae that are to be filled with new facts. Among the planned undertakings for coming seasons is a research work on the palace — temple complex discovered at the Ukok and dated near the Late Middle Ages.

Among the results of several seasons of work one can enumerate three monographs to be published (one — in press of Harvard University), a series of articles issued in Russia and abroad (in France, Japan, South Korea, Great Britain — 24).

The second item of the work at the Ukok is connected with conservation of the most valuable artifacts for future exhibition, with an elaboration on the methods of conservation and restoration. Presently, a wood conservation line has been mounted, based on Japanese equipment. The line is the only one on the territory of Russia and the Commonwealth of Independent States. Ties are maintained with the Swiss Textile Restoration Centre; a series of felt objects from the Ak-Alakha 1 site were restored at the Laboratory of the Hermitage, St. Petersburg.

To sum it up I would like to emphasize the fact that several films were shot by our colleagues from Switzerland, the U.S.A., Belgium, Japan. Some of the films highlighted the methods of works, the process of investigations of various archaeological sites at the Ukok, thus providing the scientific community of the world with information.

♦

SOME ASPECTS OF COLLABORATION BETWEEN THE ROYAL MUSEUMS OF ART AND HISTORY OF BELGIUM AND RUSSIAN COLLEAGUES IN ARCHAEOLOGICAL AND ART HISTORIAN MATTERS

F. VAN NOTEN
Royal Museums of Art and History, Brussels, Belgium

ABSTRACT. Our museums have large experience in performing archaeological excavation in Belgium and in other countries. Four years ago we became interested in Russia and especially in Siberia. We are conducting excavations in two regions: around Lake Baikal and in the Altai mountains (the Ukok plateau). We proposed to sign agreements with three large Russian museums: State Armory Museum in the Kremlin, Historical Museum of the Kremlin, and the Hermitage. At present we are planning to organize at the royal Museum a new exhibition hall "Russian Museum" in which temporary exhibitions seen as joint ventures with Russian museums will be organized.

Keywords: archaeology / museums / exhibitions /pazyryk / paleolithic.

The Royal Museums of Art and History consist of six museums, all situated in Brussels. They represent the largest museum complex in the capital of Europe, with seventy museums. Belgium is a small country with the population of less than 10 million people, and it has 650 museums!

The Royal Museums of Art and History host collections from prehistoric times up to the present day, from all over the world, except black Africa. This part of the world is represented in the Royal Museum of Central Africa at Tervuren. Our museums have large experience in performing archaeological excavations in Belgium and other countries. The list of countries in which we conducted field research include Egypt, Syria, Jordan, Greece, Mexico and Peru.

Four years ago we became interested in Russia and especially Siberia. Presently, we are conducting excavations in two regions: in the Lake Baikal area (in Buryatia in 1993), and in the Altai mountains, on the Ukok plateau.

The Lake Baikal area, like the area where the Denisova Cave is situated, is of our particular interest because we are very much involved in what we like to call, culturally speaking, "modern man". It seems, indeed, that modern man (Homo sapiens) physically originated in Africa but that, in becoming Homo sapiens sapiens, man developed a series of new prehistoric cultures, which appeared as a new "cultural evolution" in Siberia. In fact, we see the earliest of these modern looking industries in this part of the world being earlier than in Western Europe. This fascinating period of about 40 thousand years might well have been the era during which the real modern man's evolution made him invent new ways to adapt to changing climatic and environmental conditions. After this crucial time, when he replaced the

V. A. Koptyug and J. Klerkx (eds.), Science Policy: New Mechanisms for Scientific Collaboration between East and West, 223–224.

Neanderthal man, his migration to the West and to the East, from Central Siberia, seems to be a fascinating topic of research in which we want to collaborate with our Russian colleagues. If we are very much interested in prehistoric industries of that time, and of course in studying the evolution of human behaviour, we are looking forward to more intensive research of modern man himself. Only then it will be possible to get involved in detailed genetic studies of how human populations are historically related to each other.

As to our investigations in Altai, I would like to refer to the articles published in Altaica and devoted to the research and excavations in collaboration with our Russian colleagues at the Ukok plateau. We are planning to continue this work for several years, contributing financially to this research. We sincerely hope to find a frozen burial in the next season.

I would like to greet here Natasha Polosmak who made the fascinating discovery of a frozen lady this season.

Being the biggest museum in Belgium, we have of course, a long-standing collaboration with large Russian museums. We have been exchanging exhibitions with the Hermitage for several years. Recently we had an exhibition of silver of the Sassanids from the Hermitage. We, in exchange, gave the Hermitage an exhibition of bronzes from Luristan. Since the Russians are interested in what we have in our museums in the field of West European art and history, we have agreed to organize a big exhibition with our treasures at the Hermitage.

In 1991 we received an exhibition on the Scythians from them. It is at that moment that the idea of collaborating with the Russians in Altai on the history of the Scythian people came to us.

Another collaboration of great importance proposed to the Russians is to continue the exchange of temporary exhibitions as well as permanent to the benefit of both sides. We have created in Brussels, in view of closer collaboration, a new gallery called "The Russian Museum." We proposed to sign agreements with three large Russian museums: the State Armory Museum in the Kremlin, the Historical Museum of the Kremlin, and, of course, the Hermitage. The idea is to organize in Brussels each year, for periods as long as 6 months, a large-scale exhibition from all of these museums, in turn. The first will take place next year: the Armory Museum will send an exhibition on the hunts and parades of the tsars. The following year the History Museum will send an exhibition, then the Hermitage.

Technically, this is not a true exchange, in which both sides send exhibits to their respective counterpart. The new idea I have proposed to my Russian colleagues is to plan these exhibitions rather as joint ventures. So, there will be no exchanges, collections will be coming only one way, from Russia to Belgium. But we would pay for the event: transport, insurance; we construct the exhibition and we print the catalogues, with the help of our Russian colleagues. The idea of joint venture is to share the eventual financial benefit. This, I think, is a new idea. If it works, it will contribute to the financial well-being of both Russian and Belgian institutes.

Furthermore, we are presently working on the idea of having a large exhibition on Shamanism in Belgium. Here again main contributions should be made by Russian and Siberian museums. These contributions will be complemented by joint Belgian-Siberian work in the field. After being shown in Belgium, such an exhibition could eventually be displayed in other countries.

III. INFORMATION SUPPORT AND EDUCATION

INFORMATION SERVICE FOR SCIENTIFIC PROGRAMS IN THE SIBERIAN BRANCH OF THE RUSSIAN ACADEMY OF SCIENCES

B.S. ELEPOV, L.K. BOBROV, O.L. LAVRIK, S.V. MOROSOV and E.B. SOBOLEVA
State Public Scientific and Technical Library of SB RAS, Novosibirsk, Russia

ABSTRACT. The article characterizies structure, conventional and electronic information resources and service forms of information system within the SB RAS. Special attention is paid to the development of a data transmission network to provide access to remote databases. The authors conclude that SB RAS possesses rather a firm information base to join in the world information community. Its regional information and library system is technologically ready to provide information for international scientific programs.

Keywords: information resources / library holdings / regional data bases / information exchange / network access / remote data bases / reviews / analytical information / specialized information centers.

Works Organization and Information and Library Holdings

The development of an information and library sphere in the Siberian Branch of RAS is determined by the peculiar conditions of SB operation. They are:

- a great variety of scientific investigation;
- a hierarchical management system;
- coordination of complex scientific programs.

These caused the formation of a hierarchical 3-level structure aimed at the best use of information resources (Fig. 1).

At the top of the structure is the State Public Scientific and Technical Library (SPSTL), which provides information for the basic scientific investigation of the Branch, its activity on complex and international programs. Information and library services of the scientific centers, within the SB, deal with the same tasks on their level. The libraries and information bodies of research institutes provide information for laboratories and researchers, making use of information resources of higher levels.

Taking into account the changes in the organizational forms of scientific investigations, e.g. the foundation of Russian and International Research Centers, the SB together with traditional libraries and information bodies organize special problem-oriented information centers. Among them are: the Library and Information Center on chemical compounds spectra, the Center on chemical aspects of environmental science and others. Such a center on Baikal problems is being established. Here books, periodicals, reference books, etc. as well as information

V. A. Koptyug and J. Klerkx (eds.), Science Policy: New Mechanisms for Scientific Collaboration between East and West, 227–234.

sources — traditional and electronic, document and factrographic ones — on a definite subject are collected.

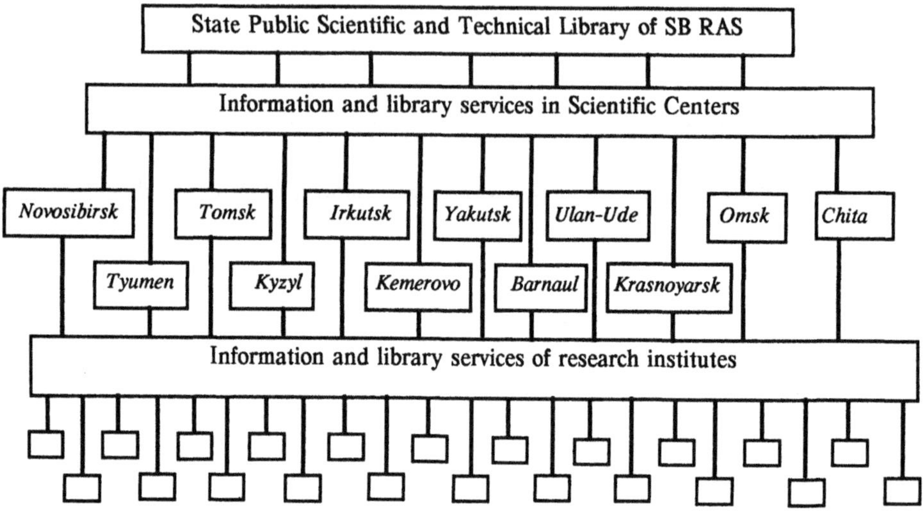

Fig. 1.
Hierarchial 3-level structure of information and library services.

Information resources to support all variety of pure and applied investigations include traditional and electronic ones. SPSTL is the only library in Siberia, obtaining a compulsory copy of all printed publications. About 90% of foreign literature in Siberia are in its stocks. The foreign publications are obtained by currency and with the help of international book exchange.

Annually, up to 1991 the Library bought 2200 foreign journals and magazines and 8000—10000 monographs. Since 1991 this quantity has sharply fallen. SPSTL exchanges books with 435 organizations from 35 countries. International interlibrary book exchange is held with 25 borrowers from 10 countries. Annually SPSTL gets near 150 requests on different publications and makes requests on 25—50 publications.

A great help to SPSTL in foreign literature acquisition is rendered by foreign organizations: American Chemical Society, English Royal Society, Soros Foundation, German Ministry of Culture and some others.

The total library holdings are more than 13 million books, periodicals, etc. Their structure is shown in Fig. 2.

Electronic information resources include databases on magnetic tapes, obtained from central bodies of State automated system of scientific and technical information; databases accessible through remote on-line searching, organized by the Centralized System of Automated Information Exchange (CSAIE); databases on compact disks, regional databases generated in the Library (Tab. 1—5).

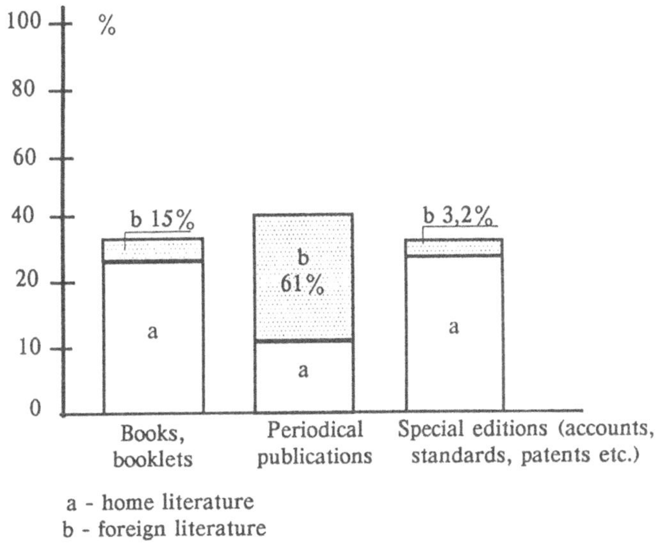

Fig. 2.
The structure of total library holdings.

Table 1. DATA BASES

* *Automation and telemechanics*
* *Computers and computer systems*
* *Semiconductor devices*
* *Materials for electronics*
* *Technical cybernetics*
* *Optoelectronic devices*
* *Software*
* *Radiolocation. Radionavigation. TV-technique*
* *Quantum electronics. Cryoelectronics. Holography*
* *Fibro-optical systems*
* *Methodology and measuring technique*
* *Biology*
* *Geology*
* *Geophysics*
* *Mining*
* *Mechanical engineering*
* *Metallurgy*
* *Mechanics*
* *Nature protection and restoration of natural resousces*
* *Transport*
* *Chemistry*
* *Economics*
* *Electrotechnics*
* *Electrical power*
* *Physics*
* *Interbranch information*

Table 2. REGIONAL DATA BASES GENERATED in the LIBRARY

* *Soils, flora and fauna of Siberia and the Far East*
* *Climate and Hydrology of Siberia and the Far East*
* *Geology of Siberia and the Far East*
* *The Nature Protection in Siberia and the Far East*
* *The Problems of the North*
* *The History of Siberia and the Far East*
* *Literature, Art in Siberia and the Far East*
* *National Economy in Siberia and the Far East*

Table 3. FOREIGN DATA BASES on OPTICAL COMPACT-DISKS

* "Computer Library" - *Computers and Programming*
* "PollTox", "Environment Library" et al. - *Environmental science*
* *"INIS" - Nuclear physics*
* "BIOSIS", "Biotechnology Abstr." et al. - *Biology*
* "Medline" et al. - *Medicine*
* *Catalogs of Publisher's houses and libraries*
* "Science Citation Index" - *Data base*

Table 4. FOREIGN DATA BASES on DISKETTES (Current contens)

* *Natural sciences*
* *Physics, Chemistry, Geosciences*
* *Social sciences*
* *Medicine*
* *Technical sciences*
* *Biology, Agriculture and Nature protection*

Table 5. FOREIGN DATA BASES ACCESSIBLE by STN-SYSTEM

* *Chemistry*
* *Engineering and applied scienses*
* *Biology*
* *Agriculture*
* *Medicine*
* *Computers*
* *Physics*
* *Informatics*

The total information volume annually processed in the system has reached 1 million documents, and the volume of computer memory, occupied by stored retrospective databases exceeds 10 Gbytes.

The access to remote native and foreign databases by CSAIE operates since 1982 [1]. Since that time the co-work of SPSTL and the Computing Center of SB RAS has resulted in the improvement of hardware, software and technology of the terminal station. One of the stages of such a development was the organization of a Training and Demonstration Center of STN-system (in the Novosibirsk Institute of Organic Chemistry (NIOC), providing access to more than 200 databases. Fig. 3 is a simplified scheme of the latter, which illustrates the technical aspect of the access to databases within the STN-system.

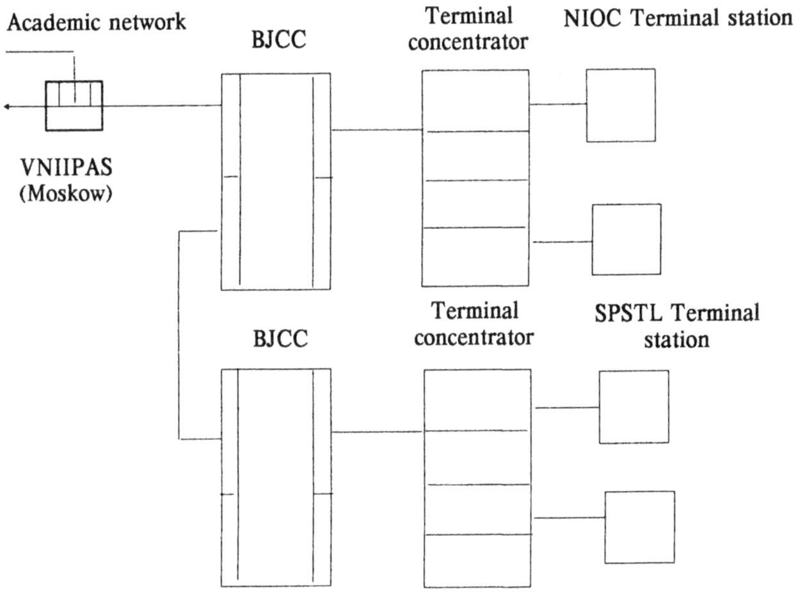

BJCC - Batched jobs commutation center

Fig. 3.
The scheme of remote access to foreign databases.

New principal tasks for SPSTL are to carry over the stored databases to compact disks and to develop CD-ROM technology in SB RAS. The solution of these tasks will advance the information service for fundamental investigation in SB RAS on a higher level.

Data transmission networks as the basis for information substructure of SB RAS

The task to form a global information community implies making use of modern tools of electronic communications. They are data transmission networks of different countries, unlike in topology, data transmission protocols, united in a system (Fig. 4).

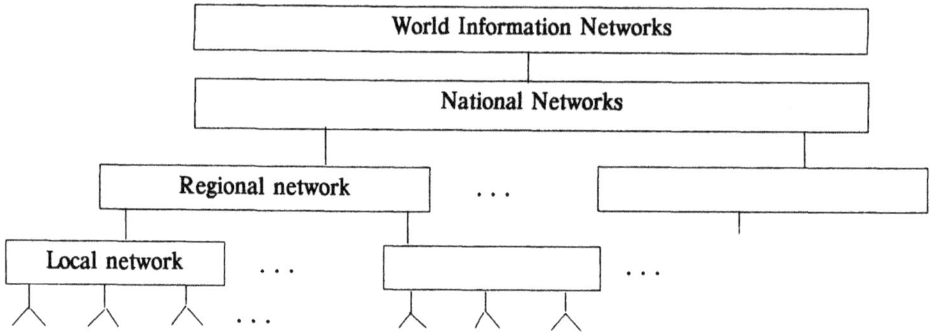

Fig. 4. The structure of networks.

The major tasks of local networks as a source of an ascending document stream are information acquisition and its initial processing (it does not matter what — bibliographical or factorgaphic ones) and transmission up to the regional level (regional networks). The tasks of this level are to systematize, process and aggregate the data obtained, to form databases and banks with this regional information and to transmit it up to the national level for further analysis and global decision-making.

If to consider how users get information, it is easy to notice that they can be conditionally divided into 4 levels (Fig. 9). The first and a very impressive one is the users of the local network; the second level is users with a regional (branch) status, whose number is rather great too, as, roughly speaking, the users of interbranch host-centers and networks can also be attributed to this group. Users of the national and world networks are relatively few, although the rank of those users and level of their possibilities (i.e. decision-making) are undoubtedly high.

In practice the role of the regional level can be played by the branch one, and a general structure of integrated networks may not have a definite hierarchical character. Nevertheless, in any case it is the local and regional networks and systems that are the basis for a communication system, as it is they that are more closely connected with information resources and users.

Thus, the urgent development of local and regional data transmission networks is a necessary condition for creating a modern information substructure in Siberia.

The experience gained at Automated System of Scientific and Technical Information (ASSTI) operation demonstrates urgent need to form an information network of the SB, within which a remote access to different databases of regional ASSTI should become possible, as well as participation in teleconferences, and the use of E-mail. To fulfill this task the following work is being done:

- realization of a network access to information resources of national centers of scientific and technical information;
- organization of terminal access to teleconferences system and E-mail;

— the use of Novosibirsk-Moscow network for the access to remote electronic information resources;
— perfecting the technology of E-mail use for solving information and library tasks (to order literature by interlibrary loan, to correspond with partners on international book exchange, information service in the "request — answer" mode, etc.);
— the formation of the system for the integral databases use.

The solution of these tasks assumes the formation of flexible multifunctional communication system, able to incorporate different terminals and to provide a common use of both remote and local databases.

The Forms of Information Service and Formation of Specialized Information and Library Centers

Within the frame of the regional information and library system created by SPSTL the following forms of information service for scientists and specialists are possible:
— selective dissemination of information (i.e. awareness news) using databases;
— retrospective information retrieval in local, remote databases as well as that in compact optical disks;
— formation of problem-oriented databases according to user's orders, able to operate within an information retrieval system;
— supply of literature (information) lists and bulletins of new books in the SPSTL holdings using the SPSTL electronic catalogue by on-site access or from remote sites;
— formation (together with organizations interested) of specialized information and library systems and centers on actual scientific problems.

In the last case the system consists of 3 interconnected compounds:
— inquiry file for primary sources in the specialized library provided with an electronic catalogue;
— bibliographic databases;
— technological solutions on analytical use of primary sources and databases.

The integration of the abovementioned information allows the realization of a closed cycle of information service:

an information need → a request → retrieval and search in bibliographic and/or factographical databases → analysis of the retrieval results → search for requested primary sources in the electron catalogue → primary sources analysis and synthesis of reviews, forecasts, etc → decision-making.

During 1990—1992 the SPSTL together with NIOC carried out work necessary to form information service on chemical aspects of environmental protection with the following basic tasks: acquisition, analysis and information supply on the problem "Chemistry and sustainable development".

This work resulted in the formation of the basic version of an integral system for research and applied works information supply.

Review and Analytical Work

Despite the variety of printed (traditional) and electronic home and foreign publications it is rather difficult to use them for solving specific scientific and technical problems. That's why it is an analytical review that becomes rather popular information and scientific product, and a necessary element for any problem solution.

Initiating scientific programs on environmental protection the Siberian Branch of the RAS has organized the information service which includes analytical reviews. They are united in the series "Ecology", which is being published by SPSTL since 1989 [2]. This series is for researchers, technologists and decision-makers, dealing with pure, applied and social problems of environmental science.

The main attention in the 27 reviews published was paid to the problems of toxic chemicals spreading, determination and neutralization, methods for toxicants analysis; anthropogenous impact of industrial enterprises, electrotechnical systems, transport on man and environment; industrial and municipal wastes treatment and management; non-waste technologies; environmental law.

The surveys are being written by the leading specialists of the RAS and other academic and applied institutes and industrial enterprises. They are those, who investigate that very problem and, hence, can and are to add their own research results to data available in literature. These results may be unique as they are based on the investigation of regional environmental problems or on the practical experience gained at regional enterprises and it may be their first publication. It is this information that can interest the world scientific community.

Computer-based technology for a review preparation gives an opportunity to get and offer:

— published reviews (500—800 copies);
— printer copies of reviews on diskettes;
— the text of a review or its parts on diskettes or paper;
— the lists of referred literature.

Information service makes use of possibilities of regional ASSTI covering some dozens of SB institutions.

Conclusion

The Siberian Branch of the RAS possesses rather a firm information base to join in the world information community. The regional information and library system of SB RAS is technologically ready to provide information for international scientific programs.

A dramatic economic situation is reflected in financing information and library substructure of SB, and hence, in its development. One of the possible alternatives toward stabilization is to incorporate SPSTL as a SB center of scientific and technical information into purposely financed international projects and programs.

References

[1] Bobrov L.K., Karnachuk V.I., Soboleva E.B. (1985). Operation organization of CSAIE terminal user station within the frame of information service technology. Automatized Information and Library systems. Novosibirsk, pp. 14—26. (In Russian).

[2] Lebedeva A.N., Lavrik O.L. (1993). Review and analytical information on chemical aspects of environment science. J. Ecol. Chem., № 1, pp. 51—53. (In Russian).

♦

STATE AND PERSPECTIVES OF COOPERATION OF THE INSTITUTES OF SIBERIAN BRANCH OF RUSSIAN ACADEMY OF SCIENCES WITH WESTERN SPECIALISTS IN THE FIELD OF GEOPHYSICAL INFORMATICS

A.S. ALEKSEEV
Siberian International Centre of New Information Technologies,
Computing Centre of SB RAS, Novosibirsk, Russia

ABSTRACT. The paper considers the potential of the Siberian Branch, the Russian Academy of Sciences, in the fields of applied and computational mathematics, mathematical modelling in the problems of studying natural resources, processes and disasters, such as earthquakes, tsunami, large meteorites and asteroids.

Keywords: mathematical modelling / natural resources / natural phenomena / disasters / multidisciplinary approaches.

Practically all the institutes of the Siberian Branch of the Russian Academy of Sciences (SB RAS) take part in the scientific development of the problems of search, study, exploration and rational use of Siberian natural resources: oil, gas, ore minerals, diamonds, coal, forests, land and water resources, etc. These complex problems as well as the development of fundamental science, creation of scientific-educational and scientific-technical infrastructures in Siberia are the basic problems of the Siberian Branch.

Investigations in ecology, global changes of the environment and climate, natural catastrophes were intensified in recent years. Work is being conducted to obtain geological-geophysical knowledge, and in aerospace monitoring and digital cartography of the territory of Siberia that remains a "white spot" on the global geophysical maps.

Mathematical methods of geophysics, mathematical modelling on the basis of the multidisciplinary approach play an important role in complex problems of natural processes and resources.

The scientific potential of the Siberian Branch in the field of mathematical modelling of different fundamental and applied investigations is high and is based on scientific schools on applied and computer mathematics that exist in Siberia and are still developing here. The leaders of these schools were the academicians M.A. Lavrentiev, S.L. Sobolev, S.A. Khristianovich, G.I. Marchuk, L.V. Kantorovich, N.N. Yanenko, A.P. Ershov. The unique, world-renown Russian schools of P.L. Chebyshev, S.A. Chaplygin, A.M. Lyapunov, M.V. Ostrogradsky, A.N. Kolmogorov, A.N. Krylov, M.V. Keldysh, A.N. Tikhonov are the origin of these schools. The scientific potential of these schools, realized in the system of uni-

V A Koptyug and J Klerkx (eds), Science Policy New Mechanisms for Scientific Collaboration between East and West, 235–245

versity education and in all scientific groups of Siberia having high mathematical background, is the means promoting the development of methodology of multidisciplinary mathematical models and informational-computational methods necessary for our contemporary society. The potential of the Siberian Branch of Russian Academy of Sciences becomes strategically very important.

1. Improvement of Methods of Digital Data Processing in Geological Prospecting

Seismic prospecting is the most effective method in searching for oil and gas deposits in Western Siberia. It helped to detect, investigate and prepare to bore about 80% of all deposits. Owing to superhigh productivity, economic feasibility and effectiveness of the methods of seismic prospecting it was possible to create a powerful oil-gas complex in Western Siberia in 15—20 years. This is a unique geopolitical phenomenon of the end of this century.

Theoretical results of several Russian geophysicists played an important role in the development of methods of seismic prospecting. These are the diffractional transformation method of Yu.V. Timoshin (a later modification of the method of D. Claerbout that is well-known now as seismic field "migration") and the ray method of calculation of seismic fields (its authors are V.M. Babich, A.S. Alekseev, B.Ya. Gelchinsky). These results formed a theoretical basis for the digital processing methods and interpretation in the western and domestic systems of computer processing.

Layered models of the geological medium were sufficient for Tyumen conditions that are complex from the geological and geographical viewpoints (swamps, regular sedimentation).

However, in Eastern Siberia where the geological structure of the medium is much more complex (trapped fields in the upper part, permafrost zones, complex relief, etc.) such improvements of the mathematical models and processes were required as the elastic model, nonhomogeneously block complexes in the upper part, etc.

New methods of wave modelling in three-dimensional media were developed. They were based on special highly-effective numerical methods. The Professor of Stanford University John Claerbout observed that "we have no supercomputers and therefore we had to develop superalgorithms". These methods make it possible to solve the same problems with the use of the old PC versions as the problems solved by the American geophysicists on the Cray computer in several hours. Figure 1 shows an example of models of an elastic medium of the Layered block type in the presence of traps (the layer is moving at the speed of $V_p = 6$ km/sec).

Unique algorithms for the introduction of corrections into the recorded fields for anomalous structure of the upper part of the geological cross-sections have also been developed.

These results were obtained at the Computing Center of SB RAS in cooperation with the trust *"Sibneftegeofizika"*. A combined package of computer data processing for complex conditions and for small deposits is being developed now with the firm *Discovery Bay* from Houston using these results. The President of this firm is T. Mitchell and Vice-President is E. Lichman. This firm is developing the system part of the package.

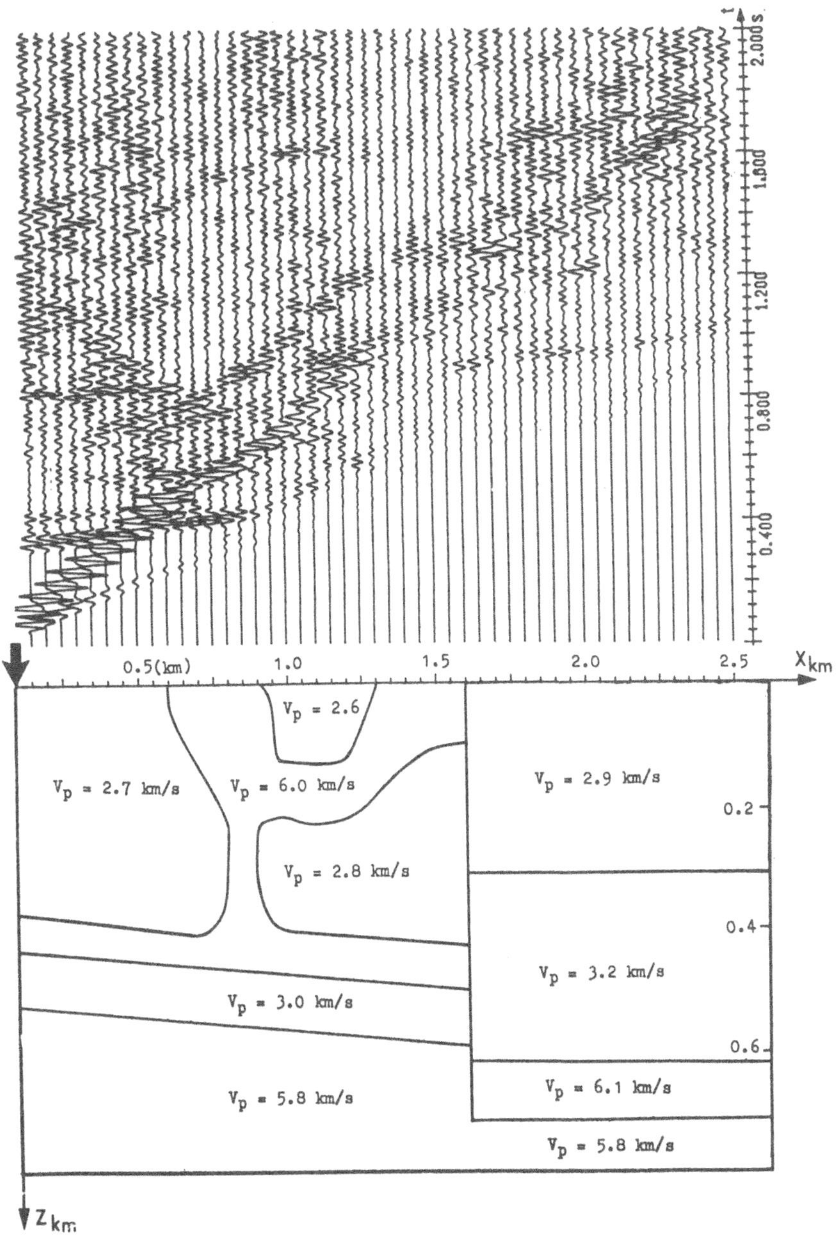

Fig. 1.
A model of the medium and a seismic shot section for the component of vertical displacement (the transverse velocity is determined from $V_s = V_p/1.73$).

2. Seismicity Study and Models of Earthquake Prediction

Southern Siberia is the most seismically dangerous area of Russia. It borders on the areas of high seismicity of Middle Asia, Mongolia and China. Works on the study of seismicity have been constantly conducted in the Computing Center of SB RAS, in the Institute of Geology and Geophysics, Altai-Sayan Expedition, Institute of the Earth's Crust, Baikal Seismic Expedition. Also, new methods for the calculation of synthetic seismograms have been developed and dynamic mechanisms in the sources have been studied.

In the last two years Chinese seismologists began to issue the results of observations for the last 25 years in the Russian-Chinese journal *"Earthquake Prediction"* that is being published in the English language. These results were obtained from a dense multidisciplinary network of stations and can serve as an empirical basis for the construction of the multidisciplinary physical-mathematical model of combined precursors of earthquakes. Such models are being developed by Siberian geophysicists and mathematicians at the Computing Center of SB RAS and the Institute of Mathematics of SB RAS on the basis of combined inverse problems of several geophysical methods such as seismic, gravimetric, electrometric, hydrological methods.

A project of international cooperation of Russian, American, Chinese and Japanese seismologists was prepared on the basis of theoretical developments on this model. The purpose of this project is testing and improvement of the model using mathematical experiments with the real measurements accumulated for many years in these countries.

Earthquakes still remain the most dangerous natural catastrophes. Earthquake prediction is one of scientific problems that should have high priority. As there are differences between intraplate and interplate earthquakes, it is necessary to begin international cooperation for the development of theoretical and computational aspects of prediction. Intraplate types of earthquakes prevail on the Eurasian continent and interplate type — in the North and South America, on the shores of the Pacific Ocean and along the median-oceanic ridges of the Atlantic, the Indian and North Oceans.

As for the central part of Eurasia, i.e., the Baikal zone, two types of earthquakes — intraplate and interplate — are available here due to the presence of growing rift structure. Therefore the Baikal zone is the most suitable zone for the study of complex problems of earthquake prediction. Moreover, the three-dimensional structure of the Earth's crust and mantle of this zone has been well studied by the methods of deep seismic sounding, seismology and prospecting geophysics developed at the Institute of Geology and Geophysics of SB RAS, the Institute of the Earth's Crust, Novosibirsk Vibroexpedition, the Computing Center of SB RAS and the organizations of the Committee of Geology of the Russian Federation. A test site for the vibroseismic monitoring and seismic tomography of seismic prone areas exist for two years already in the Baikal zone. All this, besides the International Centre for the Study of New Tectonics located here serve as a unique basis for the development of works on the earthquake prediction.

3. Vibrational Geotechnologies

Works on vibroseismic sounding of the Earth have been conducted in the Siberian Branch since 1977 based on cooperation between the Computing Center of SB RAS, Institute of Geology and

Geophysics of SB RAS, Institute of Hydrodynamics of SB RAS and Novosibirsk Vibroexpedition and design institutes such as the Institute of Applied Geophysics and the Institute of Computer Engineering.

A number of experimental models of stationary and mobile vibrators have been created. Effective pumping of seismic energy using mechanical sources at medium frequencies was realized. A method of continuous (up to several hours) accumulation of energy of signals during their recording was developed. Signals can be transmitted to the distance of as long as 2 thousand kilometers. A broadband impulse seismogram was obtained for distances up to 315 km. This seems to be a record for the Guinness book.

Figure 2 shows a fragment of a test site located near Novosibirsk. It has vertical and horizontal hydroresonant sources. They were quickly mounted *in situ* from tubes with a large diameter, pneumo-compressors and a PC computer.

Fig. 2.
This is a fragment of the vibroseismic test site near the village Bystrovka. A vertical hydroresonant vibrator and a horizontal vibrator (at the stage of construction) are shown.

Figure 3 shows a vibroseismic network on Lake Baikal with a powerful vibrator in the settlement Babushkino and with stationary and mobile stations. The range of sounding is approximately 150 km, and the depth of sounding is up to 100 km.

This experiment on sounding is similar to the experiment of the American geophysicist McEvilly from Berkeley. On Baikal, however, information for a larger territory (150×150 km^2) is being controlled and undergoes multidisciplinary processing. Here the processes of evolutionary and catastrophic development of seismicity can be observed continuously taking into account migration of zones of accumulation of stresses and changes of rheological parameters of these zones.

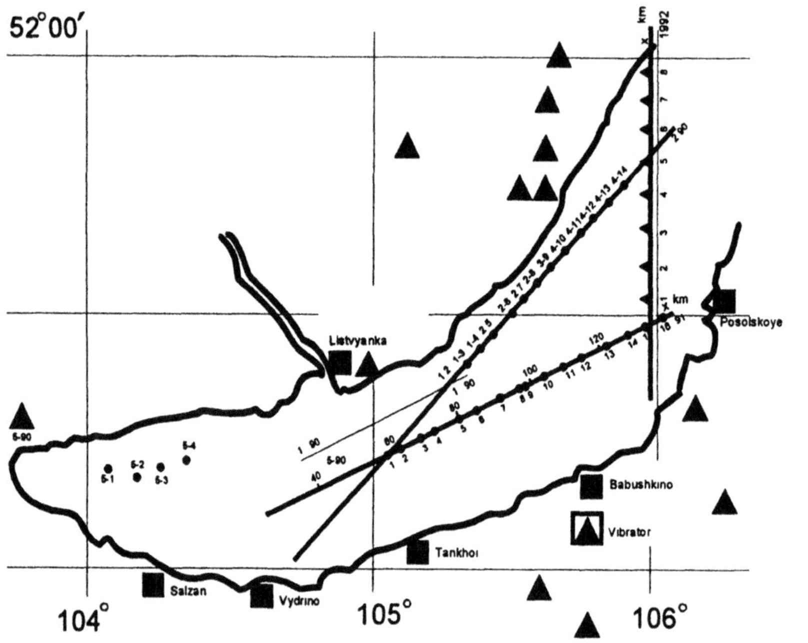

Fig. 3.
Location of profiles of deep seismic sounding, continuous seismic profiling and vibroseismic monitoring in the southern part of Lake Baikal.
Locations of bottom seismic stations are denoted by dots.
Populated localities are indicated by squares, and seismic observation points — by triangles.

International cooperation on the study of vibroseismic monitoring of seismic prone zones and multidisciplinary approaches to earthquake prediction at the Baikal test site would have good scientific perspectives. The grant on the "Active Vibroseismic Monitoring of Seismic Prone Zones and Areas of Volcanism" was given by the National Science Foundation of the USA for these works. New vibrational geotechnologies can be developed (Appendix 1).

4. Study of Meteorite Craters (Astroblems) in Siberia

Data on the impact structures of cosmogenic origin on the territory of Siberia, Far East, China and India is not complete on the world map. From the total amount of about 150 astroblems on the territory of Siberia, China and India verified by geological methods there are hardly 10 craters. This shows that these territories are insufficiently studied, if we take into account their sizes and equal probabilities of falling of celestial bodies on territories having equal areas. It is supposed that there are several hundreds of large astroblems at the bottom of oceans. By now, only 2 or 3 of them have been detected.

Investigations on the development of morphological structure and complex geophysical criteria for the identification of ring structures by remote sensing in order to understand their cosmogenic origin are being conducted in the Siberian Branch. Mathematical modelling of geological-geophysical processes (craters, earthquakes, tsunami, volcanism, dust and smoke pollution of the atmosphere, etc.) occurring when large celestial bodies fall onto the Earth's surface or into the ocean is being also carried out. Aerospace imagery of the Siberian territory showed a great amount of unstudied craters. There is a high probability that these are structures of cosmogenic origin.

Mapping of astroblems on the territory of Siberia using aerospace, geophysical and geological observations is very perspective for international cooperation.

5. Geoinformation Networks of Siberia

Materials of different expeditions, investigation and international centers were presented at the conference in the reports of the Siberian Branch. There are very many such investigation centers in Siberia now. Numerous observations are being carried out there. The state of development of processing methods of observations will make it possible to create complex mathematical models of natural processes. However, the development of telecommunication systems for the collection of multidisciplinary information for the territory of Siberia and the creation of specialized and complex analytical centers lags behind.

Some institutes function as specialized informational centers. However, transmission of information into the international sphere of processing and international exchange are difficult. Development of up-to-date technologies of collection, processing and storage of data from networks and international exchange of data is necessary here.

Many institutes have a lot of experience in the development of inter-computer and informational networks. So, beginning from 1975, 3 computer networks (a star-shaped network of remote terminals for the complex of 3 computers BESM-6, a regional dedicated digital network of Computing Center of Common Use (CCCU) for the access to ES-computers, and an ETHERNET network for CCCU for small computers) were developed at the Computing Center of SB RAS.

Then a fragment of the network IASNET (Computing Center, Institute of Nuclear Physics, Institute of Organic Chemistry of SB RAS) and a node of the network RELCOM (Institute of Information Systems of SB RAS), were created in the Novosibirsk scientific center.

Having such experience in the development of network telecommunications, the Siberian Branch can cooperate with western specialists in the creation of the Siberian sector of global informational computer networks including satellite and ground-based channels. Taking into account the necessity of informational integration of the Siberian Branch and its observational stations, analytical centers and local networks into global scientific networks, it is reasonable to develop an international project of the Siberian Geoinformation Network (Appendix 2).

Appendix 1

NEW GEOTECHNOLOGIES USING ACTION OF POWERFUL VIBRATORS ON THE GEOLOGICAL MEDIUM

Advantages

A new direction in geotechnologies exists at Novosibirsk Computing Center and some other institutes of Siberian Branch of Russian Academy of Sciences Powerful seismic vibrators that are new in design, instruments of control, recording and processing of information on the basis of PC AT, methods of mathematical modelling of vibroseismic wave fields in complex geological structures have been developed here

Application field

These developments are the basis of new geotechnologies in the following areas

- vibrational effect on oil-bearing strata for the increase of the oil yield,
- vibrational stimulation of the process of ore leaching,
- prevention of mountain shocks and unexpected discharges of coal and gas in mines (and in future, possibly, of earthquakes) by their controlled initiation and gradual discharge of stress,
- cardinal solution of the problem of prompt monitoring of the tensely deformed state in earthquake prone zones of the Earth's crust through their periodic sounding by vibrational signals This can lead to the development of scientific approaches to the prediction of time of occurrence of earthquakes,
- better microzoning of the areas where large industrial objects (e g atomic power stations) will be built through the imitation of earthquakes by powerful vibroseismic sources,
- detection of dangerous resonance properties of engineering constructions for the updating of projects and timely prediction of their possible destruction during earthquakes,
- high-resolution seismic prospecting of oil and gas deposits (seismic stratigraphy) which is based on the extension of a band of seismic signals towards both the low (4 to 6 Hz) and high (300 to 500 Hz) frequencies Realization of the methods of complex quantitative geological-geophysical interpretation

The Computing Center has all necessary conditions for the realization of the technologies listed above patents and "know-how" for powerful hydroresonant sources, highly-effective algorithms and programs of mathematical modelling of geological media allowing the discovery of new types of elastic waves (State Register of Inventions No 402, 1991), development of instrumental systems for the realization of new technologies asserted by certificates of authorship

---

Address
Academician A S Alekseev, Director
Computing Center, Lavrentiev Prospect, 6,
Novosibirsk 630090, Russia
phone (7)(383-2) 35-56-50
fax (7)(383-2) 32-42-59
e-mail aleks@comcen nsk su

Appendix 2

THE PROPOSAL FOR COLLABORATION ON THE DEVELOPMENT OF SIBERIAN GEOINFORMATIONAL NETWORK (SIB GIS NET)

Introduction

A large volume of information on natural resources and ecological state of the environment in Siberia has by present been accumulated by institutes of the Siberian Branch (SB) of RAS and the universities of Siberia

The Siberian scientific community has already started, on this basis, the activities in international research and personnel training projects

The efficiency of the collaboration in this field can be essentially increased by entering the Western geoscience databases

This proposal formulates the major objectives and initial steps in this direction

Objectives of the SIB GIS NET Project

1 Inclusion of the geoinformation data of the Siberian territory into the World's system databases

2 Adaptation of current information on natural resources and ecological state of the environment in Siberia to international standards

3 Development of a draft project on adaptation of existing dataware, software and hardware for SIB GIS NET

4 Development of legal, financial and technological approaches for the access of SIB GIS NET to the international computer networks

5 Informational support for the participation of Siberian scientists in international research and training programmes and projects in the area of the Earth's sciences

Starting point of the project

It is proposed to discuss the project, prepared by the Russian side, with representatives of the interested international organizations (UNESCO, CIFEG, etc) and national centers of the countries – members of NATO and to hold the relevant workshop in the city of Novosibirsk in April – May, 1994

Basic participants From the Russian side

- Novosibirsk Computing Center (SB RAS)
- Novosibirsk University
- Altai University
- Forestry Institute (SB RAS)
- Siberian International Center of New Information Technologies

Appendix 3

Performance of the multicomputer system "SIBERIA"

Computer System	Total Elapsed Time (sec)	CPU Time, ES-1066 (sec)	Speedup
ES-1066	1509	1509.00	1.0
ES-1066 + 1 PE	361	6.17	4.18
ES-1066 + 2 PE	207	6.19	7.28
ES-1066 + 4 PE	115	6.35	13.16

Table 1. Frequency Filtration
for the Image – Data Array 880 × 1310

Computer System	Total Elapsed Time (sec)	CPU Time, ES-1066 (sec)	Speedup
ES-1066	604.0	604.0	1
ES-1066 + 1 PE	25.0	0.3	24
ES-1066 + 2 PE	12.6	0.3	48
ES-1066 + 4 PE	6.5	0.4	93

Table 2. Quantum Chaos Problem

Computer System	Total Elapsed Time (sec)	CPU Time, ES-1066 (sec)	Speedup
ES-1066	240.0	240.0	1
ES-1066 + 1 PE	48.0	6.1	5

Table 3. Spectral and Spectral Finite Difference Methods
for the Computation of Synthetic Seismograms
for 2D Inhomogeneous Media

Computer System	Total Elapsed Time (sec)	CPU Time, ES-1066 (sec)	Speedup
ES-1066 + 1 PE	14421.0	61.0	1
ES-1066 + 4 PE	5484.0	68.0	2.7

Table 4. Spectral (Spectral Finite Difference) Method
for the Computation of Synthetic Seismograms and Snapshots

Parallel computer systems of high efficiency for data processing can be created within the framework of this project on the basis of the technology of block assembling using standard western integrated circuits. So, experimental models of working stations on the basis of PC and microcircuits Intel-860 are being produced at the Institute of Automation and Electrometry of SB RAS. The project of the Computing Center of SB RAS on the assembling of parallel computers of cluster type on the basis of PC and dozens of IC Intel-586 (Pentium) using domestic parallel translators and languages was approved by three expert commissions of the Russian Academy of Sciences and the Ministry of Science and Technical Policy of the Russian Federation. A similar computer system named "Siberia" using elements produced in our country was developed at the Computing Center of SB RAS in 1986 and had high productivity for that time (Appendix 3). Cluster parallel computers will have power approximately 1 Gigaflop and low cost (of about 50 million roubles) due to the availability of software produced in our country that is almost ready for exploitation.

These computers can be used as productive servers in the local PC network (with the technology of programming for networks). They can provide the solution of large complex problems of mathematical modelling in analytical centers and institutes where highly productive computations are necessary.

♦

INTERACTION OF EDUCATION AND SCIENCE IN SB RAS AND INTERNATIONAL COLLABORATION IN EDUCATION AND TRAINING OF JUNIOR SCIENTISTS

Yu.L. ERSHOV
Novosibirsk State University, Novosibirsk, Russia

ABSTRACT. The paper describes the highlights of organization of the educational process in Novosibirsk State University, the steps of the Siberian Branch of the Russian Academy of Sciences for re-establishing reliable links with other universities and stresses the importance of international cooperation in higher education and training junior scientists.

Keywords: quality of research / quality of education / international cooperation.

I would like to start my talk from the NATO newsletter, which every participant has, or, at least had, in his hands, with the very important address of Dr. Edward Donvitt, which contains the statement of the transition in the research paradigm in the United States and partly in NATO scientific policy from the paradigm of investigator-initiated research to the paradigm of goal-oriented research. I want to quote a very important statement from the conclusion of this address: "In these circumstances there is an important safeguard available, i.e. the strict maintenance of quality." This remark, if we agree with it, and I believe we agree, has some important implications. To maintain high quality of scientific research, there needs to exist at least a good system of higher education, the good universities. The good universities really need to be involved in scientific research activity. So, to maintain a high level of the university, there should be a critical mass of free scientific activity in fundamental science independent of what kind of paradigm we follow.

Fortunately, as far as the Siberian Branch of the Academy of Sciences is concerned, its organizers - the Academicians Lavrentyev, Sobolev, Khristianovich, Vekua - from the very beginning understood the importance of educational support for academic research institutes. Novosibirsk State University was established in 1959, with the principal aim of training junior scientists for working in Siberian scientific and higher educational establishments. For quality education and training students, the so-called scheme of the Moscow Physico-Technical Institute was used. But here in Novosibirsk, this scheme was realized more perfectly. Highlights of this scheme are following. Most professors, lecturers, assistants, scientific advisers in the University are researchers from academic institutes here in Academgorodok, and they have only part-time jobs in the University. The second feature is that students at the graduate level spend most of their time in research laboratories and take part in the real research which is occurring in these institutes. To have this opportunity, the curriculum was

V. A. Koptyug and J. Klerkx (eds.), Science Policy: New Mechanisms for Scientific Collaboration between East and West, 247–250.

created so that in the first three years the students take, very intensively, the most basic educational courses in their respective speciality and only special courses in the senior years.

To follow this process smoothly, there needs to be some especially smart students. In the Siberian Branch, our university maintains the whole system of searching for talented pupils from secondary schools from Siberia, the Far East and Middle Asia. Through the system of Olympiads, summer and winter schools, correspondence schools, etc., under the University, we have a special boarding seasonal school which has about 600 pupils, and this school provides good students for our University.

This educational system oriented on training junior scientists is working quite well. Here in Academgorodok, more than half of all researchers graduated from Novosibirsk University. The entire graduates class is about 17000, and about 5000 have Ph.D. degrees, and more than 300 have a D.Sc. degree — this is the second, higher scientific degree in our country. Three are academicians, i.e. full members of the Russian Academy of Sciences, 5 are corresponding members of this Academy.

This scheme works well so far, but there are at least two dangers. The first one is that for many years, a job at the University was a unique possibility for a scientist to have some extra money. Now, it is much easier to have much more money in other, commercial, enterprises, and it is a real problem. The economics faculty especially feels the real danger of it. Secondly, the general problem which was discussed here: this is the brain drain. If for research institutes the danger is only permanent emigration, and long-term trips are not so dangerous for research because the researcher will return with new knowledge and so on. But for educational process it is a real problem. So far we do not have a problem of finding good professors to take the main lecture courses in the University, but now (at least at the Faculty of Physics) they are beginning to feel this problem.

Now, there is of course a much more important danger: the danger of collapse of all our fundamental studies, fundamental research in our country, and it is very upsetting. If our government will sharply decrease financial support of fundamental science, we can meet the danger, and I even can say, it will be a disaster not only for Russian science, but a disaster for the whole world's science.

So far I've spoken only about Novosibirsk State University, but actually, the Siberian Branch of the Academy of Sciences understands the importance of maintaining quality in the whole system of education and takes some new steps in its support. I will just indicate two points. For example, the scientists from higher educational establishments are invited for participation in many complex scientific programs which are supported by the Siberian Branch of the Academy of Sciences, for example the national program "Siberia". And the second, maybe a formal but an important step in the creation of better relations between academic sciences and science in higher-educational institutes was the following decision of our Branch of Academy of Sciences. It was a policy of the former Ministry of Higher Education and Science to create some regional scientific-educational complexes, simply to overcome the separation of our science into academic and educational science. Such complexes were created in Altai (Barnaul), Krasnoyarsk, Omsk, Kemerovo and a few other places. And now heads of such complexes formally and really became members of the Presidium of the Siberian Branch of the Academy of Sciences.

Now, I want to turn to the second topic of discussion concerning international collaboration in education and training junior scientists. I would like to return to the starting sentence about the genuine importance of maintaining the quality of research and therefore of maintaining the quality of education and training. International collaboration in this field is extremely important for being at a good international level. For example, yesterday in Dr. Kulipanov's presentation, it was indicated that student participation in his international center is quite efficient and quite important for students to rise more quickly.

It is important for educational institutes or universities simply to have the permanent possibility of comparing and checking their curriculums with those in the universities all over the world helping each other and so on. Also, for students, international contacts are very important for widening their general view of all the problems. Novosibirsk State University has more than 30 bilateral agreements with other universities. We have agreements with some American universities and with universities in Germany, Italy, China, and many other countries. A good exchange has been going on for years with Stanford University. It started from just a people-to-people exchange at students level, but we believe that we are going to the next higher level of connection and some of our students in physics now are post-graduate students at Stanford. Our Physico-Mathematical school has a good permanent (now, 6-year-old) exchange program with Phillips Academy in the United States, a well known elite boarding school in America, e.g., the former United States President Bush graduated from that Academy. Now, almost all of these agreements do not work for different reasons, but the main of these reasons is the financial situation in this country, because the usual scheme of such agreements envisages a currency-free exchange, and yesterday the Minister spoke about this scheme. And now it is not working well, and so we must find some other possibility. We have good cooperation with Oldenburg University in Germany, it is working well in different directions, and this university really understands our problems. Last year it decided to spend all the money which they have for two years on cooperation with our university, just on our science to pay for tickets for our faculty and our students who are going there.

I cannot say that we can objectively evaluate our own place in the world educational system, but education in basic natural sciences in the Novosibirsk State University meets the highest international standards in mathematics, physics, chemistry, biology, geology and geophysics, linguistics, history and so on. At present, because of the changes in the general situation and political situation in this country, we have to change some of our curriculums in economics, humanitarian sciences, and here it is quite important that we are helped by western universities. I can say that the University of Oldenburg and the University of Paris IX help us to create a new curriculum in economics and in some other subjects. Also we have a good working agreement with the International Institute of Christian Studies in the United States, which helps us in establishing a new Faculty of Philosophy.

For some of new directions of education, e.g. ecological education, it is important to have permanent international collaboration in creating educational programs. Last year in the University a higher Ecological College was created, and this College promised its students that they would spend a few months in different ecological centers all over the world. This is important for education, but I do not know how it will be possible to fulfill this promise. Well, what is possible to do? It is possible and beneficial to enlarge the educational services for foreign students and researchers here in Academgorodok at the University and in our academic research center. We have begun advertising master and Ph.D. programs in mathematics, physics, chemistry and biology, and so on at the University.

If we have a sufficient amount of foreign students, or post-graduate students, there will be rather good support, at least financial support of our international collaboration, we can send our students and faculty abroad, to attend international conferences and to do joint research. And this is a good possibility, because you know that in many sciences the reputation of Siberian scientists is very high, and I think, Academgorodok is a really good place. Young researchers can spend some time here and improve their educational level.

Boris Elepov spoke about the importance of communication for real international connection in science. The Siberian International Center for New Informational Technologies was established, just to support new telecommunications equipment for educational and scientific purposes. This center was established by the Siberian Branch of the Academy of Sciences on one side, and by the State Committee for Higher Education on the other side, to house institutes and a computing center in Novosibirsk State University. This center has plans and good projects, and good ideas, from our foreign participants we have good promises, but so far we have had no real financial support of any of our quite interesting projects.

I would like to finish with one concrete proposal. Many times I attended meetings of Oberwolfach Mathematische Forschungsinstitut in Germany. This institute performs a special role. In this institute, every week, for the whole year, i.e. 52 weeks a year, international meetings on different topics in mathematics, are held. The Institute has an excellent library. The meetings involve 40 selected scientists from all over the world and young German mathematicians. These meetings are very efficient, very hopeful. As I understand, there was a similar institute in Trieste Centre of Theoretical Physics. And here in Academgorodok we have the possibility of establishing such an institute, which will support international meetings at all existing international centers, and to fulfill the needs of our institutes and universities for foreign schools, workshops, meetings, etc. There is a working plan of constructing an International Center of Novosibirsk State University. It will have all the necessary things: 42 single bedrooms, 6 suites, a cafe, rooms for meetings etc., and if it will be possible just to support such an institution for permanent international meetings, I think it will be a good thing. I want to indicate that this Mathematische Forschungsinstitut is supported by Volkswagen Concern, and if NATO will find a good opportunity to support such an institute here in Academgorodok, it will be a real impact of NATO in support of a new oriented international activity of the Siberian Branch of the Academy of Sciences, and of the Siberian Universities.

♦

STATEMENT

NATO/SB RAS Advanced Research Workshop
"Science Policy: New Mechanisms for Scientific Collaboration between East and West"

(Novosibirsk, Russia, 22—25 November, 1993)

STATEMENT

The role and responsibility of science for the humankind will significantly increase entering into the 21st century since global problems connected with the environment and development require global decisions based on scientific achievements, analysis and forecasting. Taking into account the universality and complexity of the aforementioned problems, we realize that it is necessary to consolidate and coordinate the global scientific efforts and to strengthen the international scientific collaboration in selected priority areas. These areas should be in line with the recommendations of the United Nations Conference on Environment and Development (Rio de Janeiro, June 1992), the ICSU Conference on Agenda of Science and Development into the 21st Century (Vienna, November 1991) and programmes of international scientific organizations.

In order to enhance the efficiency of scientific research a special attention should be paid to its methodological aspects, to the system analysis of existing problems, selection of priorities, inter- and multidisciplinary approaches, optimal balance between individual and collective efforts in science and its applications.

The participants of the workshop, from 14 countries, support the Statement of the Conference on the Role of Science in Rebuilding Central and Eastern Europe (Kent, UK, April 1993) and would like to attract attention to the following principle:

> "It was recognized that priority-setting among different needs of society entails difficult decisions, especially at a time of economic restructuring and financial stringency; nevertheless it is vital that governments make adequate provisions to safeguard science since it is very difficult to re-establish the necessary human capital and infrastructure once these have been dissipated".

Thus, the preservation and maintenance of scientific potential of Russia and other CIS countries is very important for the development of these countries and could be partly reached through the efficient international scientific and technological collaboration.

Taking into account a very difficult financial situation in science in many CIS countries, it is desirable to support from various financial sources the participation of scientists from these countries in the international collaboration — in various projects and programmes, related conferences, workshops, twinning of scientific institutes, etc.

Among new mechanisms stimulating the growth of the international collaboration in the changing geopolitical situation, it is recommended to use the existing goal-oriented open laboratories — International Research Centres — as regional nodes for joint global scientific undertakings on the way to sustainable development of our

V. A. Koptyug and J. Klerkx (eds.), Science Policy: New Mechanisms for Scientific Collaboration between East and West, 253–256.

civilization. In the framework of such Centres it is easier to find an optimal balance between fundamental research and applied aspects, as well as between the interests of individual scientists from various countries and collective goal-oriented efforts.

Based on this decision, the Workshop has defined a preliminary list of priority scientific interdisciplinary fields which should be investigated within the scope of the International Research Centres, involving a cooperative multidisciplinary approach:

- rational use of natural resources and sustainable development of regions;
- monitoring of the current state of ecosystems of the biosphere;
- assessment of natural and anthropogenous risks:
    a) interdisciplinary study of the environment evolution and of natural hazards;
    b) study of man-provoked risks;
- advanced technologies and material sciences;
- economic, social and cultural problems;
- networking, goal-oriented data bases and modelling;
- higher education and goal-oriented training of junior scientists.

More detailed inventory of priority areas and topics is presented in the appendix.

Improvement of the scientific information exchange has been recognized as an urgent need for supporting the scientific collaboration between East and West:

- mutual access to scientific databases should be made available for all scientists;
- missing databases should be urgently created in the aforementioned priority scientific areas;
- to improve data transfer in cooperative research projects and to allow access to large data pools, a fast (satellite supported) scientific DATA NET should be established.

The Workshop has recognized the necessity of giving priority to the involvement of young researchers (at the graduate and post-graduate level) in the research activities of the International Research Centres. Short-term fellowships should be made available to scientists from these Centres to perform research in specialized laboratories in Western countries. At the same time the Siberian Branch of the Russian Academy of Sciences should have a possibility to offer research fellowships to western scientists for performing research in the laboratories of institutes and universities related to the International Research Centres within the scope of international research projects.

Training should also involve undergraduate education, preferentially in new orientations, such as risk sciences, in the frame of international networks.

The participants of the Workshop are requesting UNESCO, ICSU, NATO Science Committee, Council of Europe, Commission of European Community and Governments of the CIS countries to take this Statement into account and pursue specific actions in the support of the scientific collaboration between East and West wherever possible.

Appendix

LIST OF PRIORITY AREAS AND TOPICS

I. Rational Use of Natural Resources and Sustainable Development of Regions

— ecological economics;
— institutional strategies of sustainable development of the Siberian region, legal aspects of using natural resources;
— sustainable development of selected (model) territories;
— land use and the agricultural reform;
— boreal forests - their role, conservation and management;
— water resources management.

II. Monitoring of the Current State of the Biosphere

— new methods and instruments for controlling priority toxicants in the environment;
— aircraft and satellite monitoring systems;
— ecological mapping;
— biodiversity monitoring and conservation;
— Lake Baikal as a natural laboratory;
— atmospheric aerosols;
— study of arid zones of Central Asia.

III. Assessment of Natural and Anthropogenous Risks

A. Interdisciplinary Study of the Environment Evolution and of Natural Hazards:

— global climate change and its consequences:
= archaeological data on paleoclimatology, man adaptation to the environment, ancient migration of population;
= record of climate changes in the lake sediments (Lake Baikal and lakes in other regions) and wood rings, etc;
= consequences of global climate changes;
— active tectonics risks, including earthquakes (regions of Lake Baikal and South Siberian mountain belt); methods of monitoring (magnetic, active vibroseismic, etc.);
— natural geochemical anomalies and their influence on ecosystems and human health;
— solar-terrestrial interaction and prediction of geophysical situation for the Eurasian region;
— molecular genetics of the ancient and modern man.

B. Study of Man-Provoked Risks

— chemical pollution risks;
— consequences of nuclear weapon tests and accidents on plants of radioactive material production and using;
— space debris problems;
— development of methodologies and techniques of forecasting the consequences of realization of large-scale industrial and agricultural projects.

IV. Advanced Technologies and Material Sciences

— mineral raw materials processing, including treatment of poor and tough ores;
— replacement of non-renewable raw materials by renewable ones;
— progressive materials for key branches of modern technique;

- material design, modelling of materials and processing;
- catalysts and catalytic processing;
- mechanochemical technologies;
- explosion, plasma, ionizing radiation, light and electromagnetic field for enhancing processes in materials preparation and modification;
- development and application of aerospace techniques;
- methods and instruments for materials characterization and standardization (composition, structure, physical properties, toxicity, etc.);
- progressive technologies of materials recycling.

V. Economic, Social and Cultural Problems

- comparison of federal state principles of West and East countries and interstate economic relations within the CIS;
- regional policy of Russia and a new role of Siberia; main directions of structural, investment and social policy of Siberian regions;
- global changes, new geopolitical role of Siberia and prospects for Siberian frontier areas;
- integration of the Siberian regions into the world economy;
- conversion of the defense industry establishments in Siberia (regional aspects and socio-economic consequences);
- sociological monitoring of the economic reform in Siberia;
- problems of demography and population migration;
- archeology and history of culture.

VI. Networking, Goal-Oriented Data Bases and Modelling

- computerized files of Russian scientific publications on selected problems;
- material properties data bases;
- implementation of a specialized documentary network;
- establishing an international information network based on the data banks for human genome analysis;
- development and implementation of regional information systems on pollution and health control for integrated sustainable environmental planning (Baikal and Altai regions);
- development of Siberian geoinformational network and data bases;
- man-made closed ecological systems;
- modelling and prognosis of natural and man-provoked disasters.

VII. Higher Education and Training of Junior Scientists

- ecological education;
- training in the area of remote access to the computerized data files of international networks.